本教材得到浙江理工大学品牌专业建设计划项目（英语）
（项目编号 dbts201603）资助

中国丝绸文化

Chinese Silk Culture

主　编　卢红君　陈　茜

副主编　杨晓玲　钱红英　施慧敏

编　者（按姓氏拼音排序）

傅　霞　干科安　鲁　瑜

裘瑜萍　魏开蔚　朱　赟

ZHEJIANG UNIVERSITY PRESS

浙江大学出版社

前言 Foreword

21 世纪大学生，作为新时代发展的见证者、奋斗者与创造者，是传播中国文化的重要力量，是中国故事的重要讲述者。因此，他们对中国文化的认知和自信尤为重要。在英语作为全球通用语的今天，用英语讲好中国故事，有助于向世界展现真实、立体、全面、生动的中国，提高国家文化软实力和中华文化影响力。《大学英语教学指南》（2017 版）指出："通过学习和使用英语……也有助于增强国家语言实力，有效传播中华文化，促进与各国人民的广泛交往，提升国家软实力。"

鉴于此，浙江理工大学依托浙江省"丝绸之府"的历史地位以及学校与丝绸的深厚渊源，组织编写了这本关于中国丝绸文化的英语教材。本教材在内容上突出基础性和文化性，适合本科二、三、四年级非英语专业学生在完成基础英语学习之后阅读，使其能够用基础术语讲述中国丝绸文化；同时，本教材也适用于面向来华留学生开设的中国文化类课程。

丝绸，起源于中国，被世界视为中国的标志之一。中国的丝绸之路对人类文明和文化沟通做出了重大贡献。本教材共分 10 个单元，内容涉及中国丝绸历史、蚕与养蚕业、丝绸种类与纺织技术、丝绸设计与服饰、丝绸艺术与工艺、著名丝绸品牌、丝绸城市与人物、丝绸之路、丝绸遗产与传承、丝绸传统与未来。本教材的主要特色有:

1. 以英语为主要语言载体介绍中国丝绸文化。目前国内有关丝绸或中国丝绸的教材以中文为主，鲜有英语教材。本教材的大部分素材来自世界著名英语报纸、杂志、网站，语言贴近现实，实用性强。

2. 教学设计严谨。每个单元的学习内容和活动均围绕同一主题展开，听、说、读、写、译并重，注重语言能力培养。每个单元都设计有导入活动、词汇学习、阅读理解和翻译，以及项目活动，同时穿插形式多样的任务活动。

3. 文化介绍与语言技能培养相结合。本教材既可以作为载体，向中国学生和留学生介绍中国丝绸文化，也可以作为专门用途英语教材，训练学生与丝绸主题有关的听、说、读、写、译技能。

4. 校本定位突出。本教材在个别单元内容的选材上，凸显浙江省及浙江理工大学与丝绸的渊源，服务于浙江省和浙江理工大学的发展定位。浙江省是丝绸与时尚产业大省；浙江理工大学的前身为浙江丝绸工学院，目前拥有浙江省丝绸与时尚文化研究中心、浙江理工大学丝绸博物馆。

本教材能够顺利出版，得益于浙江理工大学外国语学院和浙江大学出版社的大力支持，也得益于编写组全体成员的通力合作和辛勤劳动，在此深表谢意。由于编者水平有限，本教材的疏漏和错误之处，请大家批评指正。

编　者

2019 年 9 月

目录 Contents

Unit 1 A Thread Through History

Unit Guide: Silk originates in China and is a cultural symbol of the Chinese nation. Silk-making is a lengthy and complicated process which includes planting mulberry, raising silkworms, unreeling silk, making thread, designing and weaving fabric, dyeing, and embroidery. In this unit we will explore a brief history of silk, its discovery, how it spreads to the rest of the world and most importantly, its significance in Chinese culture so as to lay a foundation for the study of the later units. Silk culture is not just about the making of silk. It has a much deeper spiritual dimension as described by Guan Zhong, "When the dragon wishes to be small, it morphs into a silkworm. Yet when it wants to be big, it can fill the entire universe."

Lead-in

I Test your knowledge about silk.

1. Silk is made from _____.
   (A) very fine cotton
   (B) cocoons of silkworms
   (C) fur of goats
   (D) a plant called flax

2. _____ is the seal script of the Chinese character for "silkworm."

(A)

(B)

(C)

(D)

3. The life cycle of the silkworm starts from an egg to a ______, and then to a ______ and to an adult ______— a process of complete metamorphosis.
   (A) larva, pupa, moth
   (B) pupa, larva, moth
   (C) moth, larva, pupa
   (D) larva, moth, pupa

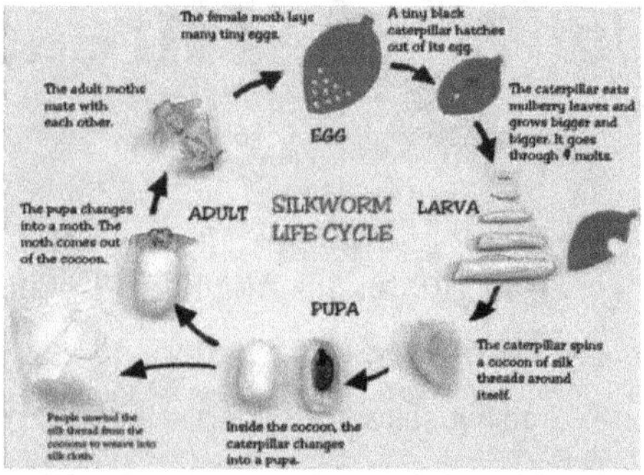

4. Silkworms are fed on ______.
   (A) silk
   (B) wheat
   (C) bamboo
   (D) mulberry leaf

5. The silkworm has to moult ______ times before it is prepared to enter the pupal phase of its lifecycle.
   (A) two
   (B) three
   (C) four
   (D) five

6. The thread of silk from a single cocoon can be ______ long.
   (A) 3 m-to-9 m
   (B) 30 m-to-90 m
   (C) 300 m-to-900 m
   (D) 3,000 m-to-9,000 m

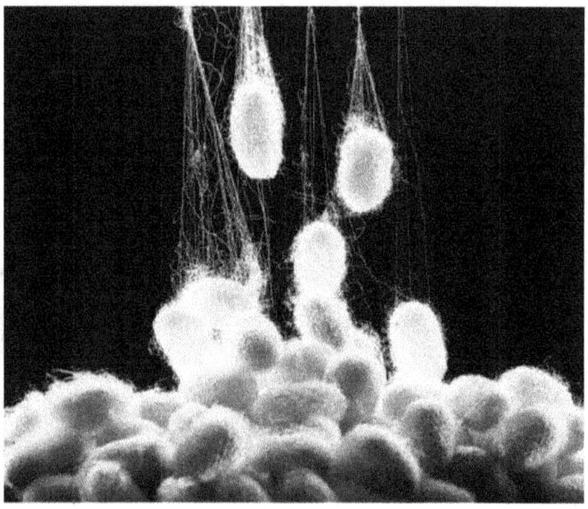

7. Legend has it that ___________ first invented the process of making silk. (A) Confucius (B) Marco Polo (C) silk dragon (D) Leizu, the wife of Yellow Emperor	
8. What kind of fibre is silk? (A) Synthetic fibre. (B) Artificial fibre. (C) Natural fibre. (D) Elastic fibre.	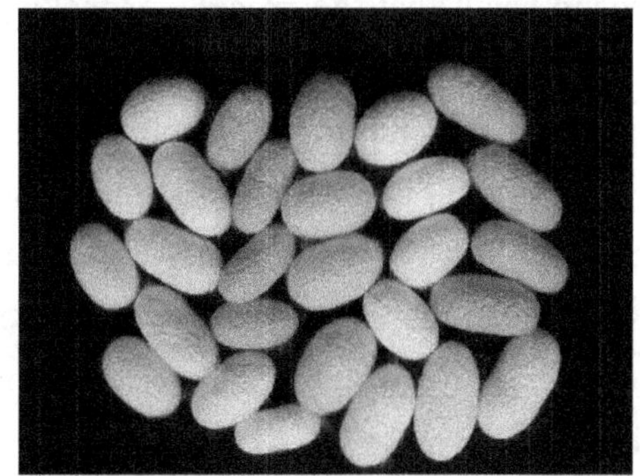
9. What could be used to loosen the fibres so they could be made into long threads? (A) Very small needles. (B) Hot water. (C) Violent shaking. (D) Soy sauce.	
10. Silk could be used ______. (A) to make clothing (B) as currency (C) as paper (D) All of the above	

II A brief introduction to silk.

Step One: Listen to the recording and fill in the blanks with the exact words you hear.

音 频

What is silk? Silk is a thin, but strong fibre that silkworms produce when they are making their cocoons. It can be 1. __________ into a very soft and smooth fabric.

Legend has it that the process for making silk cloth was first invented by the wife of the Yellow Emperor, Leizu, around the year 2. __________. The idea for silk first came to Leizu while she was having tea in the imperial gardens. A cocoon fell into her tea and 3. __________. She noticed that the cocoon was actually made from a long thread that was both 4. __________. Leizu then discovered how to combine the silk fibres into a thread. She also invented the silk 5. __________ that combined the threads into a soft cloth. Soon Leizu had a 6. __________ of mulberry trees for the silkworms to feed on and taught the rest of China how to make silk.

Silk cloth was extremely valuable in Ancient China. Wearing silk was an important 7. __________ symbol. At first, only members of the 8. __________ family were allowed to wear silk. Later, silk clothing was 9. __________ to only the noble class. Merchants and peasants were not allowed to wear silk. Silk was even used as 10. __________ during some ancient Chinese dynasties.

Step Two: A debate

In the past, the emperors of China wanted to keep the process for making silk a secret and anyone caught telling the secret or taking silkworms out of China was put to death. If you were one of the emperors, would you keep the silk-making a secret or share it with other countries?

Reasons for keeping it a secret	Reasons for sharing it with others

Ⅲ Splendid Chinese silk.

Watch the video and complete the following sentences.

视 频

1. At the G20 Hangzhou Summit, a variety of silk products could be found throughout the conference venue, such as ______________________________ ______________________.

2. Silk products presented to the foreign guests as gifts at the G20 Hangzhou Summit contain traditional Chinese cultural motifs, such as ______________________________ _____________, and they are representations of sublime craftsmanship. Such traditional charm is mixed with a touch of ______________________________, the very essence of the 5,000-year-old Chinese culture.

3. The history of silk almost developed ______________________________ the civilization of China. Via the Silk Road, Chinese silk was shipped regularly to foreign countries gaining praise for its beauty, ____.

4. In recent times, silk production underwent dramatic changes as human labour was _____ ______________________________ in silk mills.

5. Silk represents Chinese civilization and has become a beautiful ______________ for the cultural communication between the East and the West.

6. The spread of silk involved more than simple duplicating, but featured as well a ________ ______________________________.

Passage 1

A History Wrapped in Pure Silk

Zhao Xu

In 53 BC, **Marcus Licinius Crassus**, a Roman general and politician who once acted as a consul for the Roman Republic, was fighting troops of the **Parthian Empire** near the town of **Carrhae** (now Harran, Turkey). In the heat of battle, the Parthians were said to have produced a military flag made of Chinese silk that, under the **glaring** sun, **blazed resplendently**. **Dazed** by a **luminance** they had never seen before, the Romans **panicked** and were eventually defeated.

The story is just one of many about Chinese silk, a famous product once transported on a **transcontinental route** known today as the Silk Road.

"The term Silk Road was **coined** at the end of the 19th century by the German geologist and **orientalist Ferdinand von Richthofen**, who between 1868 and 1872 did extensive research in China," said Zhao Feng, director at the China National Silk Museum[1] in Hangzhou, Zhejiang Province.

Zhao pointed to a **gilt** bronze silkworm **unearthed** from Shaanxi Province, whose capital, Xi'an, once served as the capital of Han Dynasty[2]. The city was also located at the starting section of the Silk Road, first opened during the **reign** of the Han emperor Wudi[3] between 139 BC and 126 BC.

The worm is a mere 5.4 centimetres long. Its body is divided into nine segments, and includes forelegs as well as middle and hind legs. Both its **taut** body and the slightly raised head are clear indications that the worm is in the process of spitting silk. It has been dated to the Han Dynasty.

"The level of realism achieved by the **artisan** responsible for the bronze silkworm is almost startling," Zhao said. "It **testifies** not only to the **consummate craftsmanship**, which could only be achieved through repeated **rendering** of the same image, but also to the **emblematic status** the tiny worm commanded at the time."

Silk became the No. 1 **merchandise** on the Silk Road partly because it was light in weight, so

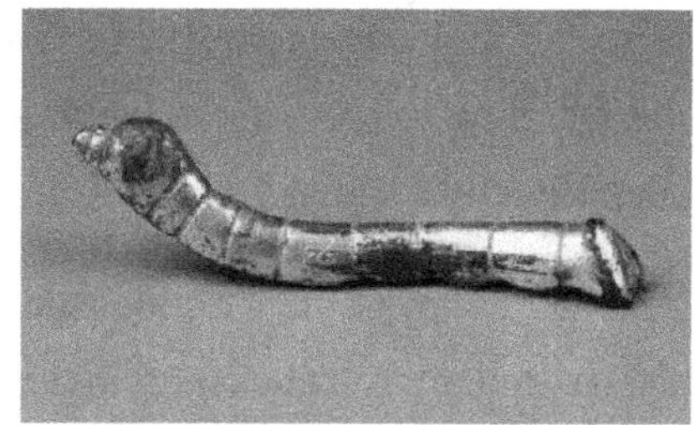

was easy to carry in relatively large quantities. Its warm reception by people far and wide not only **yielded** vast profits for the middlemen—mostly **Sogdian** merchants who routinely sold the much **sought-after** commodity at a price a hundred times higher than its **initial** purchasing cost—but also made silk the **de facto** hard currency[4] on the trading route.

Legends **abound** regarding the westward transmission of **sericulture** and silk **fabric** making techniques. One appeared in the *Records of Xiyu During the Time of Tang*[5], a twelve-volume **travelogue reputed** to have been **dictated** by Xuanzang[6], a **Buddhist** monk, who, in his **quest** for religious **scriptures**, travelled all the way from Xi'an to India and then back between 628 and 645. Most of his journey was along the Silk Road, which throughout its history had been a major **conduit** for the spread of religion.

"According to Xuanzang's book, the ruler of a Xiyu kingdom sent an **envoy** to ask for silkworm eggs from 'the country in the east' but was refused," Zhao said. "**Undeterred**, the king proposed to marry a princess from that country. This time, he was given the nod. The bride was then secretly told by the envoy that the land she would be marrying into had no silkworm. 'Bring some so that we can make your own clothes there, ' the princess was told, and **obliged**, without the knowledge of her emperor. The secret was no more."

"No date is given to the story, but we have reason to believe that 'the country in the east' refers to China, known to the ancient Greek and Roman geographers as 'Serica, ' the country **equated** with sericulture."

Judging by **archaeological excavations**, the method for raising silkworms had reached Xiyu by the third century. In one case, Han-dynasty wooden **bobbins** unearthed in the area were still wrapped around with gray and **ocher**-colored thread, **emanating** silken light after 1,800 years under the ground.

It is believed that **Nestorian** Christian monks were sent by the **Byzantine** Emperor **Justinian** Ⅰ as spies on the Silk Road from **Constantinople** to China to steal the silkworm eggs, resulting in silk production in the Mediterranean.

However, the idea of sericulture as a heavily guarded secret has always been debated. Sun Ji, a Chinese historian and **archaeologist**, believes that the Chinese government in ancient times rarely did anything to prevent the **leaking**-out of silk-making methods but, rather, had pushed for their **dissemination.**

"The main aim of the Chinese rulers in their efforts to reach out through the Silk Road was always to demonstrate the strength and greatness of their empires. Acting in a **shortsighted** and **self-interested** way would only have **undermined** this goal."

Even if they did, the fact that silk-producing was being carried out nationwide would have made it virtually impossible to keep the secret within the country's **porous** borders.

Artistic exchanges happened in due time, and the silk fabrics provided the perfect **canvas** for them. A **tapestry** unearthed in Xinjiang and dated to the Han Dynasty features a **centaur** that seems to have charged directly out of Greek **mythology**. In another case, a piece of silk **brocade**, also from Xinjiang and dated to the era of Tang, is decorated with grapevines and pairs of **facing** roosters and rams. While the facing animals constitute a signature **decorative motif** of **Sasanian Persia**, the grapevines are believed to have been growing in the abundant sunshine of the Mediterranean, before finding another home in Xinjiang, where they took root both in the sun-**drenched** soil and in local art.

Yet the most **telling** example is a silk **banner** unearthed from neighbouring Qinghai Province. Believed to have come from between the fourth century and the sixth century, the banner has on it the horse-drawn **chariot** riding Grecian sun god **Helios**. But this Helios, during his eastward journey along the Silk Road, must have been subject to a heady mix of cultural influences. As a result, his seated position is that of the **Buddha**, legs crossed, on a lotus blossom. The way the silk was woven is unmistakably Chinese.

Meanwhile, the whole silk culture in China had long taken on a deeper dimension that is more spiritual than material. One of that is best **embodied** by the image of a silkworm.

"For Chinese, the **metamorphosis** a silkworm goes through during different stages of life, and the way it breaks free of the **cocoon**, transformed, is **allegorical**," Zhao said.

"The worm, often rendered in jade or precious metals, were buried underground with our ancestors. There, it is supposed to **bestow** upon the dead the same transformative power, and to guide them into the world of **eternity**, with their magic, **glistening** thread."

A jade pendant discovered in Henan Province and dated to nearly three millennia ago is carved in the shape of a silkworm. But its head is that of a dragon, of which the Chinese consider themselves **descendants**.

A line in a book **penned** by the Chinese philosopher **Guan Zhong** says, "When the dragon wishes to be small, it **morph**s into a silkworm. Yet when it wants to be big, it can fill the entire universe."

Words

glaring [ˈgleərɪŋ] *adj.* 耀眼的；闪闪发光的

blaze [bleɪz] *vi.* 发光；照耀

resplendently [rɪˈsplendəntlɪ] *adv.* 耀眼地；灿烂地

daze [deɪz] *vt.* 使（某人）迷乱而不能做出正确反应；使茫然

luminance [ˈluːmɪnəns] *n.* 亮度；发光度

panick [ˈpænɪk] *vt.* 使恐慌

transcontinental [ˌtrænzˌkɒntɪˈnentl] *adj.* 横贯大陆的

route [ruːt] *n.* 航线；路

coin [kɔɪn] *vt.* 杜撰；创造

orientalist [ˌɔːrɪˈentəlɪst] *n.* 东方通；东方学者

gilt [gɪlt] *adj.* 镀金的；涂金的

unearth [ʌnˈɜːθ] *vt.* 出土；发掘

reign [reɪn] *n.* 君主统治时期；当政期

taut [tɔːt] *adj.* 紧的，绷紧的

artisan [ˌɑːtɪˈzæn] *n.* 技工；工匠

testify [ˈtestɪfaɪ] *vi.* 证明，证实

consummate [ˈkɒnsəmeɪt] *adj.* 至上的；圆满的

craftsmanship [ˈkrɑːftsmənʃɪp] *n.* 技术；技艺

rendering [ˈrendərɪŋ] *n.* 表演；演绎

emblematic [ˌembləˈmætɪk] *adj.* 象征性的；象征的

status [ˈsteɪtəs] *n.* 地位；身份

merchandise [ˈmɜːtʃəndaɪs] *n.* 商品；货物

yield [jiːld] *vt.* 生产；获利

sought-after [ˈsɔːtɑːftə] *adj.* 广受欢迎的

initial [ɪˈnɪʃl] *adj.* 最初的；开始的

de facto [ˌdeɪ ˈfæktəʊ] *adj.* 实际上；事实上

abound [əˈbaʊnd] *vi.* 非常多；大量存在

sericulture [ˈserɪˌkʌltʃə] *n.* 养蚕；蚕事

fabric [ˈfæbrɪk] *n.* 织物；布

travelogue [ˈtrævəlɒg] *n.* 游记

repute [rɪˈpjuːt] *vt.* 把……称为；认为

dictate [dɪkˈteɪt] *vt.* 口述；命令，指示

Buddhist [ˈbʊdɪst] *adj.* 佛教的

quest [kwest] *n.* 追求；探索

scripture [ˈskrɪptʃə(r)] *n.* 经文；圣典

conduit [ˈkɒndjuɪt] *n.* 渠道；通道

envoy [ˈenvɔɪ] *n.* 使节；外交官

undeterred [ˌʌndɪˈtɜːd] *adj.* 未被吓住的；不灰心的

obliged [əˈblaɪdʒd] *adj.* 感激的；感谢的

equate [iˈkweɪt] *vt.* 等同；使相等；相当于

archaeological [ˌɑːkɪəˈlɒdʒɪkl] *adj.* 考古学的；考古学上的

excavation [ˌekskəˈveɪʃn] *n.* 挖掘，发掘；（发掘出来的）古迹

bobbin [ˈbɒbɪn] *n.* 线轴；绕线筒

ocher [ˈəʊkə] *n.* 赭色；赭土

emanate [ˈeməneɪt] *vi.* 放射；发出；散发

archaeologist [ˌɑːkiˈɒlədʒɪst] *n.* 考古学家

leak [liːk] *vi.* 透露；（指消息、秘密等）泄密

dissemination [dɪˌsemɪˈneɪʃn] *n.* 散播

shortsighted [ˈʃɔːtˈsaɪtɪd] *adj.* 目光短浅的

self-interested [ˌselfˈɪntrəstɪd] *adj.* 利己主义的；自我本位的

undermine [ˌʌndəˈmaɪn] *v.* 逐渐削弱；使逐步减少效力

porous [ˈpɔːrəs] *adj.* 能渗透的；漏洞多的

canvas [ˈkænvəs] *n.* 帆布；油画（布）

tapestry [ˈtæpəstri] *n.* 挂毯；织锦；绣帷

centaur [ˈsentɔː(r)] *n.* （希腊神话中）半人半马怪物

mythology [mɪˈθɒlədʒi] *n.* 神话学；神话（总称）

brocade [brəˈkeɪd] *n.* 凸花纹织物；锦缎

facing [ˈfeɪsɪŋ] *n.* 饰面；覆盖（如墙壁的）表面的覆饰；（衣服的）贴边

decorative [ˈdekərətɪv] *adj.* 装饰的；装潢用的

motif [məʊˈtiːf] *n.* （文艺作品等的）主题；（音乐的）乐旨，动机；基本图案

drenched [drentʃt] *adj.* 湿透的；充满的

telling [ˈtelɪŋ] *adj.* 显著的；生动的

banner [ˈbænə(r)] *n.* 横幅；标语；旗帜

chariot [ˈtʃæriət] *n.* 敞篷双轮马车（古代用于战争或竞赛）；战车

Buddha [ˈbʊdə] *n.* 佛；佛陀；佛像

embody [ɪmˈbɒdi] *vt.* 表现；象征

metamorphosis [ˌmetəˈmɔːfəsɪs] *n.* 变形（尤指从蛹变为昆虫）

cocoon [kəˈkuːn] *n.* 茧；蚕茧

allegorical [ˌæləˈgɒrɪkl] *adj.* 寓言的；讽喻的

bestow [bɪˈstəʊ] *vt.* 赠给；授予

eternity [ɪˈtɜːnəti] *n.* 永恒；不朽

glistening [ˈglɪstnɪŋ] *adj.* 闪耀的；反光的

descendant [dɪˈsendənt] *n.* 后裔；后代

pen [pen] *vt.* 写

morph [mɔːf] *vt.* 改变；变体；变形

Proper Names

Marcus Licinius Crassus （人名）马库斯·李锡尼·克拉苏，古罗马军事家、政治家、罗马共和国末期声名显赫的罗马首富

Parthian Empire 帕提亚帝国（公元前247—公元224），又名安息帝国，是亚洲西部伊朗地区古典时期的奴隶制帝国

Carrhae （地名）卡莱

Ferdinand von Richthofen （人名）费迪南·冯·李希霍芬，德国地理学家、地质学家

Sogdian 古索格代亚纳人（居住在索格代亚纳的伊朗人）

Nestorian 聂斯托利派的（聂斯托利派是基督教的一支）

Byzantine 拜占庭帝国的；东罗马帝国的

Justinian 东罗马帝国皇帝（527—565），史称查士丁尼大帝

Constantinople （地名）君士坦丁堡（土耳其港市）

Sasanian Persia 波斯萨珊王朝

Helios 赫利俄斯，古希腊神话中的太阳神

Guan Zhong （人名）管仲，中国春秋时期法家代表人物

Notes①

1. The China National Silk Museum (CNSM), near the West Lake in Hangzhou, is one of the first national-level museums in China and the largest silk museum in the world. As the largest specialized museum on textiles in China, the main goal of CNSM is to research and conserve Chinese textile relics.
   中国丝绸博物馆位于杭州西子湖畔，是中国国家一级博物馆，也是全世界最大的丝绸博物馆。作为中国最大的纺织业专业博物馆，中国丝绸博物馆的主要目标是研究和保存中国的纺织品遗物。

2. The Han Dynasty (206 BC—220 AD) was the second imperial dynasty of China, preceded by the Qin Dynasty (221 BC—206 BC). Spanning over four centuries, the Han period is considered a golden age in Chinese history.
   汉朝（公元前206—公元220）是继秦朝（公元前221—公元前206）之后的第二个大一统王朝。汉朝分为西汉和东汉，历经四百多年，是中国历史上的强盛时期之一。

3. Han emperor Wudi, or Emperor Wu of Han (156 BC—87 BC), was the seventh emperor of the Han Dynasty of China, ruling from 141 BC to 87 BC. His reign resulted in a vast territorial expansion, a strong and centralized state, and the increasing cultural contact with Western Eurasia.
   汉武帝（公元前156年—公元前87年）是中国汉朝的第七位皇帝，公元前141年至公元前87年在位。在汉武帝的统治下，汉朝开辟了辽阔的疆域，加强了中央集权，同时与欧亚之间的文化交流也不断得到加强。

4. Hard currency is any globally traded currency that serves as a reliable and stable store of value.
   硬货币是指外汇市场上汇价稳定或看涨的货币。

5. *Records of Xiyu During the Time of Tang* is a narrative of Xuanzang's journey from Chang'an in central China to the Xiyu of Chinese historiography. Xuanzang's travels demarcate not only an important place in cross-cultural studies of China and India, but

① 本书的Notes部分采用中英文相互补充的方式（非英汉对照）对文中相应内容进行解释说明及拓展。

also cross-cultural studies throughout the globe. Bianji, a disciple of Xuanzang, spent more than one year editing the book through Xuanzang's dictation.

《大唐西域记》是由唐代玄奘口述、辩机（玄奘的一个信徒）花一年多的时间编的地理史籍，记载的是唐代时期玄奘从长安（今西安）出发亲身游历西域的所见所闻。该书不仅对中国和印度的跨文化交流研究至关重要，它也是世界跨文化交流的重要资料。

6. Xuanzang (602–664) was a Chinese Buddhist monk, scholar, traveller, and translator who travelled to India in the seventh century and described the interaction between Chinese Buddhism and Indian Buddhism during the early Tang Dynasty. His seventeen-year overland journey to India (including Nalanda), which was recorded in detail in the classic Chinese text *Records of Xiyu During The Time of Tang*, which in turn provided the inspiration for the novel *Journey to the West* written by Wu Cheng'en during the Ming Dynasty, around nine centuries after Xuanzang's death.

玄奘（602—664），唐代著名高僧、学者、旅行家和翻译家。他历时17年，从长安（今西安）出发去印度进行佛学交流，这段早唐时期中国佛学及印度佛学交流的经历被记录在《大唐西域记》里，也为明朝（大约在玄奘去世900年后）吴承恩创作《西游记》提供了灵感。

Reading Comprehension

Choose the best answer to each of the following questions or sentences.

1. According to this passage, the Romans were defeated at the battle of Carrhae because __________.

   (A) they were outnumbered by the Parthians
   (B) they were startled and scared by the Parthian military flag made of silk
   (C) their general was no match for his Parthian counterpart
   (D) the Parthians boasted more powerful military force than the Romans

2. __________ created the term Silk Road.

   (A) Guan Zhong
   (B) The Han emperor Wudi
   (C) A Germany geologist and orientalist
   (D) People who did business along the road

3. The gilt bronze silkworm unearthed from Shaanxi Province __________.
   (A) is in the process of hatching eggs
   (B) reflects the best craftsmanship at the time
   (C) was buried together with the Han emperor Wudi
   (D) indicates that Shaanxi was the starting section of the Silk Road

4. Which of the following statements is NOT true?
   (A) Silk was the only merchandise on the Silk Road.
   (B) Silk was the No. 1 commodity traded on the Silk Road.
   (C) Silk was used as currency on the Silk Road.
   (D) Silk trade generated enormous profits for the businessmen.

5. The story in the *Records of Xiyu During the Time of Tang* recorded that __________.
   (A) the ruler of a Xiyu kingdom asked in vain for silkworm eggs from the emperor of the country in the east
   (B) the ruler of a Xiyu kingdom married the princess of the country in the east because he was in ardent love with her
   (C) the emperor of the country in the east granted no permission to the marriage proposal
   (D) the princess of the country in the east brought silkworms to the Xiyu kingdom with her emperor's permission

6. According to Sun Ji, __________.
   (A) silkworm-raising techniques reached Xiyu by the third century
   (B) it was Nestorian Christian monks who stole the silkworm eggs from China to the West
   (C) China rarely did anything to keep silk-making techniques as a secret
   (D) archaeological excavations find that in the past China had porous borders

7. The silk banner unearthed from Qinghai Province indicates that __________.
   (A) it was a mixture of various cultural influences
   (B) it was woven in a uniquely Chinese style
   (C) the Chinese silk fabrics provided the perfect canvas for artistic exchange
   (D) All of the above.

8. The allegorical meaning of the silk culture can be demonstrated as follows EXCEPT __________.
   (A) the silkworm artefacts buried with the dead could lead the dead into the world of eternity

(B) the metamorphosis of the silkworm goes beyond the biological transformation, but owns spiritually transformative power.

(C) the silkworm is an incarnation of the dragon

(D) silk products have been traded transcontinentally for over two millennia

Summary

Fill in the blanks with appropriate words.

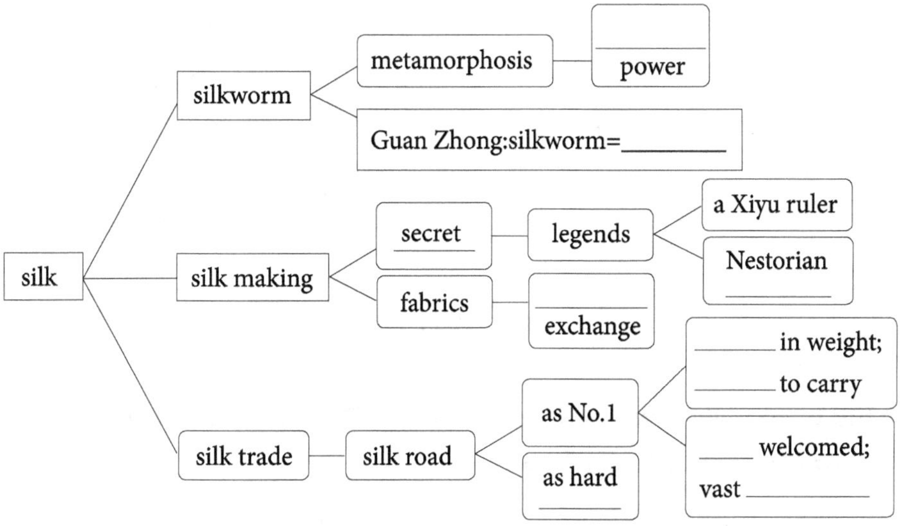

Word Building

I The adjective *fore* can be used together with other words to form new ones, meaning "before, in advance, in front of." Match the following words with their parts of speech and Chinese meanings.

	foreleg	前臂
	forehead	前景
n.	forearm	祖先
	forefoot	食指
adj.	foreground	前院
	forecast	最前列
v.	forefather	前额
	forecourt	前足
	forefinger	前腿
	forefront	预报

II The prefix *trans* can be used together with other words to form new words, meaning "across, beyond, into another place or state." Match the following words with their parts of speech and Chinese meanings.

	transcontinental	转移，迁移
	transborder	转录，抄录
n.	transatlantic	横跨大陆的
	transcribe	移植，移种
adj.	transfer	交易，买卖
	transplant	转基因
v.	transgenic	超越，胜过
	transmit	跨越国境
	transaction	横跨大西洋的
	transcend	传播，传输

Words and Expressions in Use

I Fill in the blanks with the words given below. Change the form where necessary. Each word can be used only once.

abound	bestow	brocade	canvas	coin	dissemination
glistening	raw	route	status	testify	yield

1. Love is a __________ furnished by nature and embroidered by imagination.

2. It was also indicated that over the past two years, European agriculture suffered droughts and total output of __________ silk was declining.

3. Tujia __________ is hand-made by the local women. It is with classically simple style features, wide veins of bright colours and diverse patterns.

4. Images of Sogdian musicians and dancers __________ in Tang Dynasty tombs, testifying to a willingness of the local people to be entertained in their afterlife.

5. From the underground crypt of the Famen Temple, many Tang-dynasty incense burners and related wares have been unearthed, __________ to a populous culture that was a veritable phenomenon.

6. He has 29 years of experience making wines and medicinal alcohol, earning the __________ of national-level wine taster and winemaker.

7. Proposed by China in 2013, the Belt and Road Initiative is aimed at building a trade and infrastructure network connecting Asia with Europe and Africa through the revival of ancient trade __________.

8. Since its inception in 2013, the idea and vision of the Belt and Road Initiative have been translated into action and reality, __________ fruitful outcomes with the concerted efforts of various parties.

9. Though the term *tangzhuang* was deployed to describe the APEC jacket, there was no Chinese word in the early 2000s to denote clothing from the Han Dynasty, and the term eventually __________ was *hanfu*.

10. Fujian Province highly values the conservation and utilization of documentary heritages and currently it is actively promoting the preservation, digitalization and __________ of the maritime Silk Road documentaries.

11. Jiayuguan is well known as the "throat of China" due to its strategic position at the end of the Great Wall. The historical background has __________ the city with rich tourism resources like Jiayuguan Fort and the First Beacon Tower of the Great Wall.

12. The giant cape, worn with a __________ golden gown, took its inspiration from the painted architecture of royal palaces in Beijing. "It took us five months to weave the 20 meters of fabrics we needed for the cape, experimenting constantly with the colours and suffering many failures in between," said Bu.

II Render the following Chinese expressions into English. The first letter of each word is given.

1. 种桑　p__________　m__________　t__________

2. 养蚕　r__________　s__________

3. 吐丝　s__________　s__________

4. 佛教经文　B__________　s__________

5. 考古挖掘　a__________　e__________

6. 装饰用的挂毯 d___________ t___________

7. 炉火纯青的技艺 c___________ c___________

8. 广受欢迎的商品 s___________ m___________

9. 目光短浅的中间商 s___________ m___________

10. 散发出丝绸的光泽 e___________ s___________ l___________

Translation

Translate the following paragraph into English.

中国是丝绸的故乡，丝绸文化是中国文化的一个重要组成部分。丝绸与中国的礼仪制度、文学艺术、风土民俗等有着密切的联系。比如，帝王用丝绸彰显其权威；百官用丝绸标识其等级；文人写下咏叹丝绸的诗词；蚕农举行祭祀仪式，祈求蚕丝丰产。几千年来，中国丝绸通过“丝绸之路”传播到亚、欧、非各地。中国的丝绸以其卓越的品质、精美的花色和丰富的文化内涵闻名于世，成为中国文化的象征和东方文明的使者。

Passage 2

The History of Silk Weaving and Fabric

Silk is one of the most ancient fabrics known to humanity. According to the historical records, the material first originated in China and associated with it was a royal tale of discovery. It was from here that the fabric **got a head start** after which it unfolded to Korea, Japan, Persia and other parts via **the Silk Route** (Silk Road).

Until the 13th century, the fabric was considered only for royal and the wealthy. For several years, Italy remained the European leader of silk until France promoted its silk weaving during the 17th century.

Origin of Silk

Silk existed in China before the mid of the 3rd **millennium** BC during which just 1,000 yards, or 1 kilometre, of thread from a silkworm's cocoon, was **reeled, spun** and woven. It became a significant **contributor** to the rural economy of the nation.

As per a legend, the wife of the **legendary** Yellow Emperor Huangdi discovered silk in 27th century BC as a woven fibre. With the observation of **unravelling** cocoon that fell from a mulberry tree into her cup of tea, the empress discovered the process of obtaining **shimmering** threads. She soon came up with the **cultivation** of worms, sericulture, and **conceived** the reel and **loom**. Since then, for almost three millennia, China enjoyed the global **monopoly** for producing silk.

The Growth of Silk Production in China

The Song Dynasty accomplished the art of **Kesi**, a magnificent tapestry of silk woven with a needle acting as a **shuttle** on a small loom. Records state that the artwork was found by the Central Asia's Sogdians, which was enhanced by the **Uighurs** before Chinese went for it. Kesi stands for cut silk and is derived from the **perpendicular** space between colourful areas, caused by the threads not extending across the width.

During Han **regime**, silk weaving was the primary industry and chief export product. Early Han textiles found from Mawangdui act as the testimony of weaving development in the form of brocade, **damasks**[1], and embroidery. The next remains were usually damasks, which were finely woven with colorful patterns repeating at every 2 inches or 5 cm.

Silk Secret Revealed to the World

Almost 2,200 years ago, in 200 BC, several Chinese people migrated to Korea and revealed silk production in that country. From here, the secret gradually was revealed throughout Asia. It also reached the West via various channels. In almost next 500 years, sericulture was established in India after which it started trading with Persians who developed their rich techniques in the 12th century.

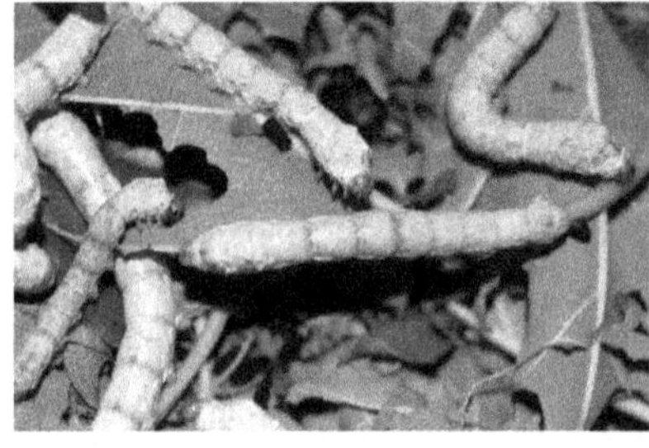

In 550, the Byzantine Emperor named Justinian got the first silkworm eggs from two Nestorian monks who had hidden them in their **concave** bamboo **staves**. The Byzantine state and church introduced imperial workshops and kept the secret to themselves due to which the silk industry boomed in the Middle East.

In the 7th century, the Arabs took over the Persians and spread sericulture throughout their victorious territories of Spain, **Sicily**, and Africa. Then the **Mongol** regime's army and Marco Polo's China trips resulted in commercial exchanges between the West and East. This led to a silk industry in Italy in the early 12th century.

During the 13th century, Italy started producing silk with the help of thousands of skilled silk weavers from the present **Istanbul** region (former Constantinople). Finally, the art of creating silk was spread into Europe.

The Improvement in Silk Technology Before Industrial Revolution

According to Chinese sources, a machine was used in 1090 to unravel cocoons that were positioned in a big hot water basin. Silk was woven onto a big **spool** due to its **back and fro** motion. While not much is known about the Chinese spinning techniques, it is recorded that a spinning wheel moved by hand was popular at the start of the Christian era. A silk spinning machine driven with a water wheel was popular in the 14th century.

According to the **compilations** of the 13th century, Chinese looms were superior of all. It stated two loom types due to which the worker's arms were set free. The first was the Eurasian **drawloom** and the second was the East Asian **pedal loom**. Since the 2nd century BC, silk brocades were formed by using **four-shafted looms**.

The Middle Ages witnessed the introduction of many techniques for manufacturing silk. While the period from 10th to 12th century saw minor changes, the 13th century marked the time of big twists. In its beginning, an ancient form of **milling** the silk threads had its hold in the industry. Then, several types of devices were introduced, whose complex structures were invented in the 14th century.

The earliest European spinning wheel together with Bobbins and **warping** machines boosted the production of silk. It is believed that the toothed **warper** was allowing a more uniform and a longer warp was first introduced by the silk industry.

The Black Death[2] incidence in end of 14th century shifted the trend towards more economical techniques. In the silk industry, water-driven mills increased in the spread and the loom set up by **Jean le Calabrais** was universally used by the 15th century.

Silk production also flourished in England in 1680s due to the **revocation** of the **Edict of Nantes**[3] that forced several skilled French weavers and experts to migrate to England to flee religious **persecution.** Likewise, silk spread to numerous nations including Mexico in 1522.

The Improvement in Silk Technology During Industrial Revolution

During the 17th and 18th centuries, further growth started for simplifying and standardising silk manufacture, most notably in the **Lyon**, France. In 1725, **Basile Bouchon** was believed to have created the first semi-automated weaving loom. In 1728, this loom was later refined by his assistant **Jean Baptiste Falcon**. In 1745 these designs were again **improved upon** to create the first fully automated weaving loom by **Jacques de Vaucanson** before being again enhanced by **Joseph-Marie Jacquard** and introduced the radical Jacquard weaving loom that processed a series of **punched** cards **in the** precise **sequence**.

These cards were a direct **antecedent** to the modern computer. Since 1801, embroidery was precisely **mechanised** because of the highly efficient Jacquard loom that also **paved the way**

Basile Bouchon's Semi-automated Loom 1725, on display at CNAM[4], Paris, France

for mass production of intricate designs.

However, in the 19th century, the rise in the price of cocoons along with the fall in importance of silk in **bourgeoisie** garments resulted in the decline of the European silk industry. However, the launch of the **Suez Canal** and the shortage of silk in France **brought down** the cost of importing Asian silk, mainly from Japan and China.

Urbanisation in Europe forced several Italian and French skilled men to leave silk production for profitable factory jobs. Further, the outbreak of silkworm diseases made silk-rearing a reduced source of income.

After Industrial Revolution

After the fall of silk in Europe, sericulture's modernisation in Japan made it the leading producer. In the early 20th century, quick mechanisation made Japan the producer of 60% of the planet's raw silk. While Italy came up from the crisis, France was stuck.

However, during the Second World War, the scene soon changed! Silk supplies were stopped from Japan due to which the Western world had to come up with **substitutes** such as nylon and other synthetic fibres.

Even post-war, silk remained a luxury product. Japan **restored** itself to be the leading exporter of silk, via its improvements in reeling and inspection technology. This continued until the 1970s after which China gained back its prominent position, while Japan saw a decline due to the increasing importance of synthetic materials.

Today, India, Japan, ROK, Thailand, China and Brazil are leading producers of silk.

Words

millennium [mɪˈleniəm] *n.* *pl.* millennia 千禧；一千年；千禧年

reel [ri:l] *vt.* 卷；绕；

vi. 蹒跚，摇晃；来回旋转；眩晕

n. 卷轴；卷筒；卷盘

spin [spɪn] *vt.* 纺（线）；吐丝，将……抽成丝

contributor [kənˈtrɪbjətə(r)] *n.* 贡献者；捐助者；投稿者

legendary [ˈledʒəndri] *adj.* 非常著名的；享有盛名的

unravel [ʌnˈrævl] *vt.* 解开；拆散；阐明

vi. 被解开；被拆散

shimmering [ˈʃɪmərɪŋ] *adj.* 闪闪发光的，发微光的

cultivation [ˌkʌltɪˈveɪʃn] *n.* 栽培；耕作；教养；（关系的）培植

conceive [kənˈsi:v] *vt.& vi.* 怀孕；构思；想像；设想；持有

loom [lu:m] *n.* 织布机，织布法

monopoly [məˈnɒpəli] *n.* 垄断；专卖；垄断者；专利品

Kesi [kəs] *n.* 缂丝

shuttle [ˈʃʌtl] *n.* （织机的）梭子；（缝纫机的）滑梭

perpendicular [ˌpɜ:pənˈdɪkjələ(r)] *adj.* 垂直的，成直角的

regime [reɪˈʒi:m] *n.* 政治制度；政权；政体

damask [ˈdæməsk] *n.* 缎子；锦缎

concave [kɒnˈkeɪv] *adj.* 凹的

stave [steɪv] *n.* 棍；棒；木柱

spool [spu:l] *n.* 【印，纺】有边筒子；短管

compilation [kɒmpɪˈleɪʃn] *n.* 编辑；编写；编辑物

drawloom [drɔ:ˈlu:m] *n.* 拉花机，手工提花机

four-shafted [fɔ:(r) -ˈʃɑ:ftɪd] *adj.* 四轴的

milling [ˈmɪlɪŋ] *adj.* 成群乱转的

warp [wɔ:p] *vt.* 弄弯，变歪；扭曲，曲解

warper [ˈwɔ:pə] *n.* 整经机，整经工

revocation [ˌrevəˈkeɪʃn] *n.* （法律等的）废止；撤回

cdict [ˈi:dɪkt] *n.* 法令；命令；敕令

persecution [ˌpə:sɪˈkju:ʃən] *n.* 困扰；苛求；迫害；残害

punch [pʌntʃ] *vt.* 打孔；用拳猛击；（用压穿器）穿孔，冲孔

antecedent [ˌæntɪ'si:dnt] *n.* 先行词；前事；前情；祖先

mechanise ['mekənaɪz] *vt.* 使（过程、工厂等）机械化

bourgeoisie [ˌbʊəʒwɑ:'zi:] *n.* 布尔乔亚；资产阶级；中产阶级

urbanisation [ˌə:bənaɪ'zeɪʃən] *n.* 城市化

substitute ['sʌbstɪtju:t] *n.* 替代物；代替者

restore [rɪ'stɔ:] *v.* 修复；归还

Useful Expressions

get a head start　走在前面，领先一步

the Silk Route　丝绸之路

as per... 按照；根据；依据

back and fro　来回

pedal loom　脚踏织布机

four-shafted loom　四轴织机

improve upon　改良；改进

in the... sequence　以……顺序

pave the way for　为……铺平道路

bring down　降（价）；把（某物、某人）抬下（楼、山）；使（某物或某人）掉下（倒下）；击败……

Proper Names

Uighur　维吾尔人；维吾尔族

Sicily　（地名）西西里岛（意大利一岛名）

Mongol　蒙古族人；蒙古族人的

Istanbul　（地名）伊斯坦布尔（土耳其西北部港市）

Jean le Calabrais　（人名）让·勒·卡拉布里

Edict of Nantes　南特敕令

Nantes　（地名）南特（法国西部港市）

Lyon　（地名）里昂（法国城市）

Basile Bouchon　（人名）巴西尔·布乔，法国发明家

Jean Baptiste Falcon （人名）让·巴普蒂斯特·福尔肯，法国发明家
Jacques de Vaucanson （人名）雅克·沃康松，法国发明家
Joseph-Marie Jacquard （人名）约瑟夫·玛丽·雅卡尔，法国发明家
Suez Canal （地名）苏伊士运河

Notes

1. Damask is a reversible figured fabric of silk, wool, linen, cotton, or synthetic fibres, with a pattern formed by weaving. Damasks are woven with one warp yarn and one weft yarn, usually with the pattern in warp-faced satin weave and the ground in weft-faced or sateen weave. Twill damasks include a twill-woven ground or pattern.
   锦缎，来自阿拉伯语，是由丝、毛、麻、棉或合成纤维制成的双面提花织物，其图案由编织而成。锦缎是用一根经纱和一根纬纱织成的，通常是用经面缎纹织成的图案，底面用纬面或缎面织成。

2. The Black Death also known as the Great Plague, the Black Plague, or simply the Plague, was one of the most devastating pandemics in human history, resulting in the deaths of an estimated 75 to 200 million people in Eurasia and peaking in Europe from 1347 to 1351. It created a series of religious, social and economic upheavals, which had profound effects on the course of European history.
   黑死病，欧洲中世纪的大瘟疫；也称the Great Plague, the Black Plague, 或简称the Plague。是人类历史上最为惨烈的大瘟疫，估计全世界有7500万至2亿人死于这场瘟疫。从1347年至1351年在欧洲最为肆虐，造成了一系列的宗教、社会和经济动荡，也由此改变了欧洲历史的进程。

3. The Edict of Nantes signed in April 1598 by King Henry Ⅳ of France, granted the Calvinist Protestants of France (also known as Huguenots) substantial rights in the nation, which was still considered essentially Catholic at the time. It marked the end of the religious wars that had afflicted France during the second half of the 16th century. The Edict was later revoked in October 1685 by Louis XⅣ, the grandson of Henry Ⅳ. It drove an exodus of Protestants and increased the hostility of Protestant nations bordering France.
   南特敕令（法语：Édit de Nantes），又称为南特诏令、南特诏书、南特诏谕，是

法国国王亨利四世在1598年4月签署颁布的一条敕令。这条敕令承认了主要宗教信仰为天主教的法国国内胡格诺教徒的信仰自由，并在法律上享有和公民同等的权利。但南特敕令于1685年10月被亨利四世的孙子路易十四废除，并驱逐了新教徒，也使得法国同相邻的新教国家之间的关系更为敌对和紧张。

4. CNAM refers to Conservatoire national des arts et métiers (National Conservatory of Arts and Crafts), a research and education institution operated by the French government.
CNAM 即 Conservatoire national des arts et métiers (National Conservatory of Arts and Crafts), 法国国立工艺美术学院，是由法国政府管理运作的一个研究和教育机构。

Reading Comprehension

I Answer the following questions according to Passage 2.

1. What is the appropriate length of the thread from a silkworm's cocoon?

2. How was Kesi, a silk tapestry woven during the Song Dynasty?

3. When did silk weaving become the primary industry and chief export product in China?

4. What was the first country outside China to make silk?

5. When did the commercial exchanges of silk between the West and East begin?

6. How did the ancient Chinese spin silk?

7. How did silk production come to flourish in England in 1680s?

8. When and why did mass production of intricate designs become possible?

9. Why did silk production come to decline in Europe in the 19th century?

10. What was the current situation of silk production in the world?

II Outline the "Discover and Spread of Silk" to the world in timeline pattern.

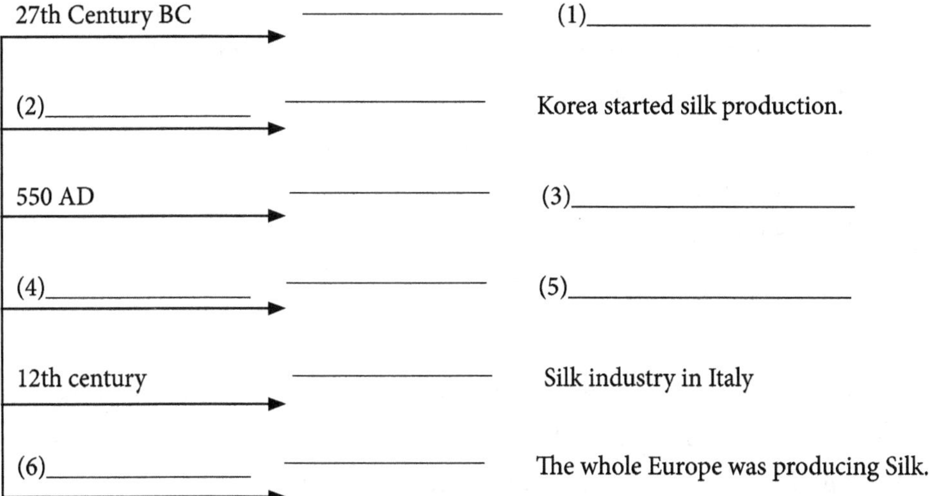

Words and Expressions in Use

I Match the collocations in Column A with the Chinese meanings in Column B.

Column A	Column B
1. reel the yarn	A. 养蚕
2. spin silk thread	B. 绕纱线
3. cultivate silkworms	C. 泄露秘密
4. conceive a loom	D. 纺丝线
5. unravel the cocoon	E. 磨平丝线
6. mill the silk thread	F. 织布
7. weave a fabric	G. 构思一台织机
8. reveal a secret	H. 解开蚕茧

II Compare the following pairs of sentences and explain the different parts of speech and meaning of italicized words.

1. reel

(A) She never finished what the man gave her to spin, and what she did spin she did not wind onto a ***reel***, but left it tangled on the bobbin.

(B) The enemy troops began to ***reel*** in defeat under our violent attack.

(C) He made a speech straight off the ***reel*** without stumbling over a word.

(D) The boy began to ***reel*** in the line slowly when he wanted to stop flying the kite.

2. spin
(A) As he lay thinking, he saw a spider over his head, trying to ***spin*** a web. He watched her as she toiled slowly and with great care.
(B) The scenario is similar to that of a figure skater drawing her arms inward during a ***spin*** to turn faster on the ice.
(C) He ***spun*** the wheel sharply and made a U turn in the middle of the road.
(D) All those figures make my poor head ***spin***.

3. shuttle
(A) The rules of engagement for the group phase of the competition read like the instruction manual for a space ***shuttle***.
(B) There's a ***shuttle*** service between the airport and the train station.
(C) Consider the simplest, plain ***shuttle*** loom on which only two harness frames are used for the shedding operation.
(D) He and colleagues have ***shuttled*** back and forth between the three capitals.

4. weave
(A) And all this beautiful silk, she said, would be used to ***weave*** colourful clouds in heaven.
(B) When completed, this lacing forms a sort of loose, springy ***weave*** that does not bind strongly.
(C) In Beijing, it seems to take forever to get anywhere as you ***weave*** your way through congested streets.
(D) They ***weave*** narratives from seemingly innocuous blogs, magazine ads, TV slots, fashion labels and public phone calls.

5. loom
(A) Another government spending crisis is ***looming*** in the United States.
(B) The spinning jenny and the power ***loom*** had made it possible to turn the cotton fibre into cloth quickly.
(C) Chinese public opinion is also beginning to ***loom*** large in a range of issues critical to Sino-U. S.
(D) When layoffs ***loom***, face time with coworkers and the boss in the office gains consequence.

6. mill

(A) On the next day, there was an attack on a woolen ***mill*** employing many Chinese workers.

(B) The figures had been circulating previously on China's rumour ***mill***, but the report by CCTV appeared to be an official confirmation.

(C) "You just have to ***mill*** it a little bit, and it reacts like ordinary cement," Stemmermann says.

(D) So many different ideas are ***milling*** about in my head, but I can't settle down to work on any one of them.

Ⅲ Fill in the blanks with the words or expressions given below. Change the form where necessary. Each word or expression can be used only once.

monopoly	flourish	mechanise	revocation	get a head start	as per...
back and fro	improve upon	substitute	restore	bring...down	
pave the way for...					

1. You already have the skills and it is usually much easier to________________ them than to start from scratch.

2. Here he sees the industry of his native country displayed in a new manner, and traces in their works the embryos of all the arts, sciences, and ingenuity which ____________ in Europe.

3. The goal was not to develop trade and expertise where none had existed, but to help ______________ Europe to its former wealth.

4. Visitors can freely switch ______________ between English and Chinese by clicking the and buttons.

5. Such entrepreneurs are often able to benefit from a lack of a state-owned ___________ in their industry.

6. I'm going to leave early and ________________ on the traffic.

7. Pilot projects such as the one conducted in Wenzhou should ___________________ larger-scale reform.

8. That pace would __________________ the unemployment rate by less than half a percentage point over the course of a year.

9. Furthermore, our products can be customized ______________ the specific requirement of our clients.

10. By 1800, science and the market economy were creating incentives and opportunities for western entrepreneurs to __________ production and tap the huge power of fossil fuels.

11. Biomass is a good ___________ for fossil energy due to its renewable, abundantly distributed characteristics.

12. Failure to follow these guidelines creates unnecessary work for other volunteers and may result in ____________ of editing privileges.

Translation Skills

翻译是一种主要靠语言作为媒介的文化交流活动，语言有意义且具有功能，当今比较流行的一种翻译标准即功能相似、意义相符。对功能的分析是对文本的宏观分析，对意义的理解和表达是翻译的核心。接下来的各个单元我们将从汉英语言的各个语言层级来讨论如何实现该翻译标准。

词语英译（1）——根据语境确定原文的词义

词是汉英两种语言转化活动中最基本的语言单位。词语意义的确定在翻译过程中十分重要。而词义对语境有很强的依赖性，同一个词在不同的语境中有不同的意义。翻译时不能只依靠词典给出的定义，而应根据具体的语境具体分析词语，才能正确地理解和表达词义。要想解决这个问题，需要平时的大量积累，了解意义相近的词在用法上的差别。请看下面的例子：

（1）得病以前，我受父母宠爱，在家里横行霸道……

这句话中的“横行霸道”的一个常见解释为“任意欺凌”。但本句是以一个孩子的口吻写的，所以这里的“横行霸道”只是表明孩子在家里“为所欲为，想干什么就干什么”，所以 get all the things my own way 才是符合该语境的译文。

（2）混合班级会影响聪明学生的发展。

“混合班级”很容易被译成mixed classes，结合整句话来看，这里的混合显然是指能力的混合，而不是其他元素的混合，所以在翻译时需要把这一点翻译清楚，译成mixed ability classes。

（3）……使中央委员会领导年轻化

这里的“使……年轻化”如果简单译成“make...younger”就会贻笑大方，显然这个“年轻化”其实表示的是让更年轻的人来担任中央委员会领导职位，即“bring younger members to the leading positions of the Central Committee”。

（4）缂丝采用“通经断纬”的方法……

缂丝是中国传统的丝织法，“通经断纬”通俗来讲就是用本色丝线做经线，竖的经线的方向是连通的，用各种彩色丝线做纬线，横的纬线的方向由操作者控制。基于对这一四字短语的理解，本句可译为“Kesi adopts a weaving method that passes through the warp and breaks the weft.”或者“Kesi adopts a weaving method known as continuous warp and discontinuous weft.”

歧义是汉语中很常见的一种现象，源于词义的丰富性和语法的独特性。如果不放在特定的语境中，可以有两种甚至两种以上的理解。例如：“借老王一本书”，这里的“借”可以是“借给”，也可以是“借走”；“我去上课”，“上课”可以是“讲课”也可以是“听课”；“自行车没有锁”，这个“锁”可以是动词，也可以是名词；“参加大会的有工人、科技人员和专家四十余人”，可以是总共四十余人，或者专家就有四十余人。英译容易引起歧义的词或词组，必须借助语境，厘清逻辑关系，分析隐含的含义，请看下面的例子。

（1）在公交车上的时候，一个爸爸在对他儿子说：“都那么大了还打不过你妈妈，我十二岁就能打过你奶奶了！”

旁边一哥们冲那爸爸吼道：“有你这么教孩子的吗？会不会当爹啊！”

那男的愣了愣，说：“我说的是羽毛球！”

这句话中，“打”的多重语义引起了歧义，分别表示“打人”和“打羽毛球”，巧的是“打”这个词在英文中对应的beat也有多种意思，使得原句的双关效果得以在译文中重现。

On a bus, a father was talking to his son, “You are old enough, however, you can't beat your mother, me, at the age of 12, I could already beat your grandma.”

On hearing that, one guy beside roared at the father, “How were you teaching your child? Unbelievable! You have no idea of being a father, hymn!”

Astonished for a short while, the father said, “I mean that I beat her at badminton!”

（2）剩女的原因有两个，一是谁都看不上，二是谁都看不上。

在这个句子里“谁”可以是“看不上”的主语，也可以是“看不上”的宾语。这句话的翻译无法像上句保留原文的韵味和效果的同时，又不产生歧义。这时候只能舍弃保留原文效果，只译出原文含义。

There are two reasons for remaining spinster. One is that the woman appreciates nobody, and the other is that nobody appreciates her.

Exercising Your Skills

Read the following pairs of sentences and point out the mistakes in the English translation and then make corrections.

1. 在中国，眼下外语文凭的含金量较其他文凭的含金量高。
   At present, a foreign language diploma contains more gold than other diplomas in China.

2. 川味火锅已风靡全国，并走向世界。
   Sichuan hotpot has become a popular dish all over China, and is walking towards the world.

3. 心有余而力不足。
   One's ability falls short of one's heart.

4. 看来我非得用我的关系不可了。
   It looks like I have to use my relations.

5. 是否让他担任这个职位需要看他的政治条件。
   Whether or not he is given the position depends on his political conditions.

Unit Project

Silk is more than just a fabric in China: It is our history, our culture, fashion and a symbol of prosperity. It is a luxury that has been sought by emperors and citizens alike—a commodity that has connected the world and launched global commerce on an unprecedented scale.

Today, silk is yet another word for elegance, and silk garments are prized for their versatility, wearability and comfort. Silk absorbs moisture, which makes it cool in the summer and warm in the winter. Because of its high absorbency, it is easily dyed in many deep colors. Silk retains its shape, drapes well, caresses the figure, and shimmers with a luster all its own. Thousands of years later, our desire for silk is no less intense.

Contemporary silk garments range from evening wear to sports wear. A silk suit can go to the office and, with a change of accessories and a blouse, transform into an elegant dinner ensemble. Silk garments can be worn for all seasons.

Work in groups to complete the following work:

1. Describe to group the first silk dress you ever had. Do you like it or not? And why?

2. Work together to Internet or library for designs of silk garments and present to the class the dress or design you vote to be best. Give the reasons for your choice.

Unit 1 Keys

Unit 2 Silkworm and Sericulture

Unit Guide: Five thousand years ago when the first cocoon fell into the Empress' teacup, no one would know that this unprepossessing worm would transform the course of a nation and the trade routes of the world. In **Unit 2**, we will explore the life cycle of the magical silkworm, the legendary origin of silk, the history of sericulture and the process of silk making in China, and most importantly how silk became the "Queen of Textiles" and how China got the name "Seres," the Silk Nation.

Lead-in

I Test your knowledge about silk.

Step One: Vocabulary learning.

Choose and write down the correct words given below in brackets.

cocoon	larva	spin the filament	silk	egg	raw silk
hatch	moth	chrysalis	mulberry leaf	moult	metamorphosis

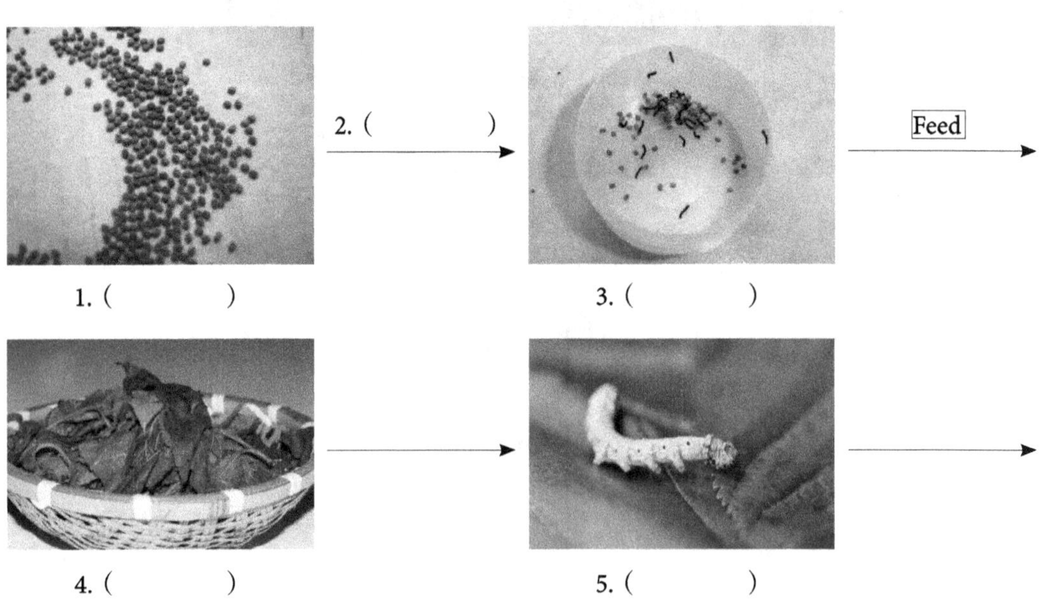

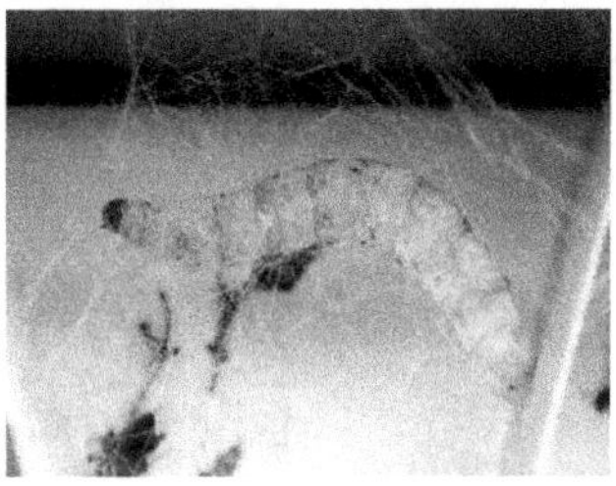

6.  (　　　　)　→　7. (　　　　)　8. (　　　　)

9.  (　　　　)　→　10. (　　　　)　→

11.  (　　　　)　→　12. (　　　　)　→

Step Two: True or false.

Decide whether the following statements are true or false. Write "T" for true and "F" for false.

1. ________ It takes 20 to 28 days for a newly hatched larva to grow into a matured one that begins to spin and cocoon.

2. ________ Female silkworm lays about 50 eggs at a time.

3. ________ Female moth lays eggs and dies after laying eggs as she does not eat anything.

4. ________ Larvae are produced in about 2 weeks from eggs at a temperature of 18 to 25 degree Celsius.

5. ________ More than 5,000 cocoons are needed to make a pound of silk.

6. ________ Today, China and Japan are the two main producers, with more than 60% of the world's annual production of sericulture.

7. ________ The female moth is smaller than the male one, and the former crawls faster.

8. ________ After the 4th moult, the silkworm begins to spin its cocoon.

II A brief introduction to silkworm and sericulture.

Watch the video and fill in the blanks with the exact words you hear.

视 频

1. In this time, it has become the ________________________ of the Chinese nation and has also ____through ______________________________.

2. Today, ________________________________ and silk customs still exist in ______________________________, in ____ _____and Chengdu in Sichuan Province.

3. The silkworm ________________________, and then it ________________________, and ____________________ before finally ____________________________. These four stages of life are ________________________.

4. Silk damask has been widely used in Chinese history for a mountain of ______________________________, and is closely related with ______________ and_______ ____.

5. Since the late 20th century, due to ____________________________________, traditional sericulture has gradually declined. As a result, various sericulture customs and traditional weaving handicrafts ________________________________.

6. We will draw up a detailed plan, secure funding and ____________________________ ________ to teach the skills and ____ ________________________ of Chinese sericulture and silk craftsmanship.

III Stages of production.

Put the following ten stages of silk production in the correct order by writing down the number in the brackets.

( ) The silk is obtained by brushing the undamaged cocoon to find the outside end of the filament.

( ) Having grown and moulted several times, the silkworm extrudes a silk fibre and forms a net to hold itself.

( ) The larvae feed on mulberry leaves.

( ) The silk moth lays 300 to 500 eggs.

( ) The silk filaments are then wound on a reel. One cocoon contains approximately 1,000 yards of silk filament. The silk at this stage is known as raw silk. One thread comprises up to 48 individual silk filaments.

( ) The silk moth eggs hatch to form larvae or caterpillars, known as silkworms.

( ) It swings itself from side to side in a figure "8" distributing the saliva that will form silk.

( ) The intact cocoons are boiled, killing the silkworm pupa.

( ) The silkworm spins approximately one mile of filament and completely encloses itself in a cocoon in about two or three days. The amount of usable quality silk in each cocoon is small. As a result, about 2,500 silkworms are required to produce a pound of raw silk.

( ) The silk solidifies when it contacts the air.

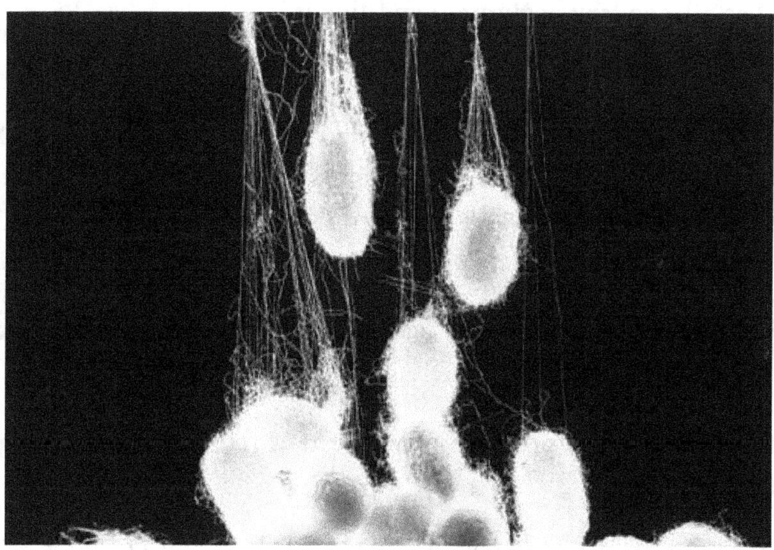

Passage 1

"Queen of Textiles" Is a Silkworm Fairy Tale

In Chinese legend, Leizu (right), the wife of Yellow Emperor, was the first to breed silkworms and make silk fabric. Thus, she was worshiped by Chinese people as the Goddess of Silkworm.

Silkworm farmers of Longzheng village in Jiangsu Province pick cocoons in the harvest season. Today, although silk is produced in many other places around the world, China remains the dominant producer.

Nearly every wonder of the ancient Chinese civilization has a legendary beginning, and this is clearly evident in silk, the "Queen of Textiles".

In ancient times, people didn't know how to make clothes out of cotton or other **refined** materials, so they covered themselves with animal hides, leaves or tree bark.

Primitive clothing **apparel** was not replaced until the arrival of the Goddess of Silkworm.

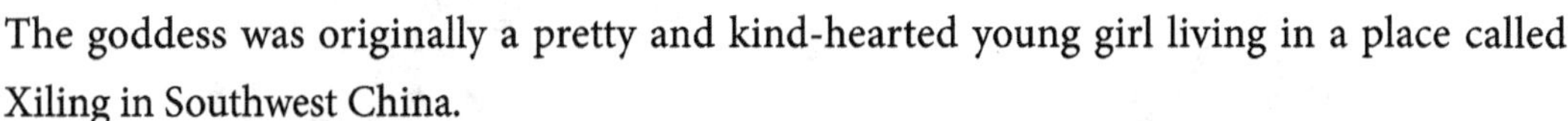

The goddess was originally a pretty and kind-hearted young girl living in a place called Xiling in Southwest China.

Unfortunately, both her parents were in ill health, so the young girl had to go out every day to gather wild fruit to feed them.

One day, she couldn't find food anywhere and worried her parents would starve. She began to cry under a **mulberry** tree. Her crying was so sad and heart-**wrenching** that Emperor Jade in Heaven[1] was touched and turned a **deity** into a worm to reside on the mulberry tree where she was crying.

The worm dropped some mulberry into the mouth of the girl, who found it both sweet and sour. When she realized that the mulberry leaf was edible, the girl was delighted and gathered some for her parents every day.

In the coming summer, the **divine** worm began to spin silk and built a cocoon for itself. The young girl then found the cocoon looking very pretty under the sunshine and started to draw out the silk and **weave** it into a piece of fabric.

When she had woven two large pieces of silk cloth, she gave them to her parents to wear. The silk cloth was very smooth and comfortable. It felt cool in summer and could keep them warm in winter.

So, the girl picked some worms from the mulberry tree and began to breed them at home. She eventually taught other people the skills of breeding silkworms and making silk fabric.

The Xiling King was very happy for this young girl's invention, so he adopted her and gave her the name "Leizu."

Later, Leizu married Yellow Emperor[2] and helped him to unite China. She also encouraged the emperor to spread sericulture (silk farming) across the country.

Ever since then, Leizu has been worshiped by Chinese people as the Goddess of Silkworm.

That's one of the legends about the origins of sericulture and silk craftsmanship in China. However, according to historical records, the origins of silk date back to 4000 BC.

In 1927, half of a silkworm cocoon was unearthed at a Xiyin Village[3] **Neolithic** Site in northern China's Shanxi Province. The cocoon was cut by a sharp knife and dated back to between 4000 BC and 3000 BC. Later, it was determined that the cocoon came from a domesticated silkworm called Bombyx Mori[4].

In 1958, **a batch of** silk **ribbons**, threads and fabric were discovered at Qianshanyang[5] site in east China's Zhejiang Province, which dated back to around 3000 BC.

In the beginning, silk clothes, and other products, were **reserve**d **exclusively** for Chinese rulers and royal families. Later, it gradually became accessible to rich people in the country. Then, silk **garments** also began to reach other regions in Asia.

However, it was not until the Silk Road opened during the latter half of the first millennium BC that the valuable fabric reached the West, where it soon became a popular luxury and the price of which was equal to that of gold.

Because of its popularity, the ancient Greeks and Romans at that time called China "Seres"[6] or the Silk Country.

Silk production is a lengthy, **laborious** and intensive process. Because silkworms are sensitive to high temperatures and picky on their food. In east China's Zhejiang and Jiangsu provinces, where there are the biggest silk producers in the country, silk can only be produced in spring and autumn each year. It usually takes around 30 days to raise silkworms from **hatching** eggs to spinning cocoons.

After being hatched, the **larvae** are scattered on large bamboo trays to grow. Since fresh mulberry leaves are the food and water of the silkworm's diet, they must be fed day and night. Also, to prevent the silkworms from getting sick, the trays must be regularly cleaned.

After four stages of growing and **moulting**, the silkworm picks a place to spin a cocoon.

Each cocoon is made of a continuous silk **filament** that is between 600 to 900 meters in length. To **unwind** or reel the filament, the cocoon has to be boiled to kill the **pupa** and then the outside end of the filament is located. As a single filament is too thin for most uses, several cocoons are reeled at the same time and with a slight twist, the filaments form a single **strand**. Several strands will then be twisted into an even thicker **yarn**.

A single silk thread **comprises of** up to 48 silk filaments and more than 2,000 cocoons are needed to produce one pound of silk.

In addition to its soft **texture** and **lustre**, silk is a natural and healthy **fibre**. It is composed of 18 **amino acid**s similar to human skin. It is capable of keeping the wearer warm in winter and cool in summer. It also breathes easily and naturally to take **moisture** away from the skin to keep it dry.

Little wonder that silk has long been **revered** as the "Queen of Textiles."

Today, although silk is produced in many other places around the world, especially in India, Japan and Brazil, China remains the dominant producer of this magic fibre. Its annual raw silk output accounts for more than 80 percent of the world's total.

Apart from garments, silk has also been used to make **quilts**, umbrellas, fans and artificial flowers. Before paper was invented, silk was used for writing and painting in China.

Because of its long history and **unparalleled** influence on not only China's but world's economy, culture and people's lives, in 2009, China's sericulture and silk craftsmanship was **inscribed on** the UNSECO'[7]s Representative List of the **Intangible** Cultural **Heritage** of **Humanity**[8].

Words

textile [ˈtekstaɪl] *n.* 纺织业；纺织品，织物

refine [rɪˈfaɪn] *vt.* 精炼；提纯；去除杂质

primitive [ˈprɪmətɪv] *adj.* 原始的；人类或动物发展早期的

apparel [əˈpærəl] *n.* 衣服，服装

mulberry [ˈmʌlbəri] *n.* 桑树；桑葚

wrench [rentʃ] *vt.* 使痛苦；使十分难过

deity [ˈdeɪəti] *n.* 神；女神

divine [dɪˈvaɪn] *adj.* 神圣的

weave [wiːv] (wove, woven) *vt.* & *vi.* （用手或机器）编，织

neolithic [ˌniːəˈlɪθɪk] *adj.* 新石器时代的

ribbon [ˈrɪbən] *n.* 丝带

reserve [rɪˈzɜːv] *vt.* 拥有；保持；保留（某种权利）

exclusively [ɪkˈskluːsɪvlɪ] *adv.* 排他地；独占地；专有地

garment [ˈgɑːmənt] *n.* 服装；衣服

laborious [ləˈbɔːriəs] *adj.* 费力的；辛苦的

hatch [hætʃ] *vt.* & *vi.* 孵化；孵出

larva [ˈlɑːvə](*pl.* larvae [ˈlɑːviː]) *n.* 幼虫；幼体

moult [məʊlt] *vt.* & *vi.* 脱皮

filament [ˈfɪləmənt] *n.* 细丝；丝状物

unwind [ˌʌnˈwaɪnd] *vt.* & *vi.* 解开；展开

pupa [ˈpjuːpə] *n.* 蛹

strand [strænd] *n.* （线、绳、金属线、毛发等的）股，缕

yarn [jɑːn] *n.* 纱；纱线

texture [ˈtekstʃə(r)] *n.* 质地；手感

lustre [ˈlʌstə] *n.* 光泽

fibre [ˈfaɪbə] *n.* （织物的）质地；纤维

moisture [ˈmɔɪstʃə(r)] *n.* 水分

revere [rɪˈvɪə(r)] *vt.* 尊敬；崇敬；敬重

quilt [kwɪlt] *n.* 被子

unparalleled [ʌnˈpærəleld] *adj.* 无比的；无双的；空前的；绝无仅有的

intangible [ɪnˈtændʒəbl] *adj.* 无形的

heritage ['herɪtɪdʒ] *n.* 遗产

humanity [hjuː'mænəti] *n.* 人类

Useful Expressions

a batch of 一批

comprise of 由……组成；由……构成

inscribe sth. on sth. 写入；题；刻

Proper Name

amino acid 氨基酸

Notes

1. The Jade Emperor is known by many names, including Heavenly Grandfather (天公), which originally meant "Heavenly Duke", which is used by commoners, the Jade Lord, the Highest Emperor, Great Emperor of Jade (玉皇上帝或玉皇大帝). The Jade Emperor is the Taoist ruler of Heaven and all realms of existence below including that of Man and Hell according to a version of Taoist mythology. He is one of the most important gods of the Chinese traditional religion pantheon.
玉皇大帝是道教神话传说中的天地的主宰，又称"太上开天执符御历含真体道昊天玉皇上帝""玉皇大天尊""高天上圣大慈仁者玉皇大天尊玄穹高上帝""玄穹高上帝""天公""老天爷"。玉皇大帝在天界的地位犹如人间的皇帝，上掌三十六天，下握七十二地，掌管神、仙、佛、人间、妖魔、地府等的一切事，权力无边。

2. Yellow Emperor, also known as the Yellow God or the Yellow Lord, or simply by his Chinese name Huangdi, is a deity in Chinese religion, one of the legendary Chinese sovereigns and culture heroes included among the Five Emperors.
黄帝：古华夏部落联盟首领，中国远古时代华夏民族的共主，五帝之首，被尊为中华"人文初祖"。

3. Xiyin Village （西阴村）is located in Shanxi（山西）Province, where half of a cocoon was discovered in 1927 to prove there existed sericulture 6,000 years ago.
西阴村遗址位于山西省运城市夏县尉郭乡西阴村的西北部，1927年在遗址中发现了半个蚕茧，证明了远在6000年前这一带就出现了植桑、养蚕业。

4. Bombyx mori (Laitin: “silkworm of the mulberry tree”) is the larva or caterpillar of the silkworm which was domesticated from the wild silkworm Bombyx mandarina.
家蚕（拉丁文：Bombyx Mori）由古代栖息于桑树的原始蚕（拉丁文：Bombyx mandarina）驯化而来。

5. Qianshanyang site is located in Lu Village in Huzhou, Zhejiang Province, which is an important site of ancient culture in the silk civilization.
钱山漾遗址，位于浙江省湖州市的潞村古村落，是人类丝绸文明史上重要的一个古文化遗址。

6. “Seres” (Laitin: “of silk”) is a name given to the silk countries or the people from these countries, especially China and Chinese people by ancient Greeks and Romans.
赛里斯（拉丁文：Seres, Sinae, Serica），意为丝国、丝国人，是古希腊和古罗马人对与丝绸相关的国家和民族的称呼。拉丁文Seres原意是“有关丝的”，一般被认为是源于中国字“丝”。

7. UNSECO: The United Nations Educational, Scientific and Cultural Organization is a specialized agency of the United Nations (UN) based in Paris. Its declared purpose is to contribute to peace and security by promoting international collaboration through educational, scientific, and cultural reforms in order to increase universal respect for justice, the rule of law, and human rights along with fundamental freedom proclaimed in the Charter of the United Nations.

联合国教育、科学及文化组织（简称：联合国教科文组织）总部设在法国首都巴黎，是联合国的专门机构。该组织旨在通过教育、科学和文化促进各国合作，对世界和平和安全作出贡献，以增进对正义、法治、人权以及《联合国宪章》所倡导的基本自由的普遍尊重。

8. An intangible cultural heritage (ICH) is a practice, representation, expression, knowledge, or skill, as well as the instrument, object, artefact and cultural space that are considered by UNESCO to be part of a place’s cultural heritage. Intangible cultural heritage is

considered by Memberas of UNESCO in relation to the tangible World Heritage focusing on intangible aspects of culture.

非物质文化遗产是一种实践、表演、表现形式、知识体系和技能，以及被联合国教科文组织视为一个地方文化遗产一部分的工具、实物、工艺品和文化场所。联合国教科文组织成员考虑将非物质文化遗产与有形世界遗产联系起来，重点放在文化的非物质方面。

Reading Comprehension

Answer the following questions according to passage 1.

1. What are the characteristics of silk cloth?

2. What did the ancient Greeks and Romans call China?

3. Why is silk production a lengthy, laborious and intensive process?

4. How can the silkworms be fed?

5. How is the filament unwound and reeled?

6. How is a thicker yarn formed?

7. What are the properties of silk?

8. What is silk reputed as?

9. What are the usages of silk?

10. Why was China's sericulture and silk craftsmanship inscribed on the UNSECO's Representative List of the Intangible Cultural Heritage of Humanity in 2009?

11. Why is Leizu revered as the "Goddess of Silkworm" ?

12. Do you know other legends about Chinese silkworm and sericulture?

Summary

The following is a summary of Passage 1. Fill in the blanks with appropriate words.

In Chinese legend, Leizu who was a 1.________________ girl was the first to 2.______________________. Thus, she was worshiped by Chinese people as 3.______________________. That's one of the legends about the 4.____________________________ in China. However, according to 5.________________, the origins of silk 6. _____________ 4000 BC. Silk production is a 7.____________________ process. In addition to its 8.________________, silk is a 9._______________ fibre. Little wonder that silk has long been revered as the "10.__________________." Apart from garments, silk has also been used to make 11. ____________________________. Before paper was invented, silk was used for writing and painting in China. Because of its long history and 12._______________ on not only China's but world's economy, culture and people's lives, in 2009, China's sericulture and silk craftsmanship was 13.______________ the UNSECO's Representative List of the 14.________________________________.

Word Building

I The suffix *ive* can be used together with other words to form new adjectives, meaning "relating to, having the quality of, tending to," and to form new nouns, meaning "a kind of person, something or abstract nouns." This suffix also has other forms like *-ative* and *-itive*. Write down the correct words in the following table.

abuse	
	cohesive
addict	
affirm	
appreciate	
	definitive
prime	
detect	
represent	

	executive
explode	
initiate	
	alternative

II The prefix *un* can be used together with other words, especially nouns or verbs to form new verbs, meaning "opposite of the original action, open or discover sth." Match the words in the left with those in the right to from appropriate collocations.

unearth	the carpet
unlock	a ball of string
unwind	buried treasures
uncover	the door
undress	the baby
unload	a suitcase
unpack	a plot
unroll	the cargo

Words and Expressions in Use

I Fill in the blanks with the words given below. Change the form where necessary. Each word can be used only once.

legend	refine	origin	weave	domesticate	exclusive
spin	unwind	paralleled	inscribe	craftsman	pick

1. He ____________ his scarf from his neck when he entered the classroom.

2. This product is adopted and passed the silkworm sand with ________ special craft as the packing material.

3. The book has enjoyed a success ____________ in recent publishing history.

4. This cloth was____________ by hand.

5. Some people are very __________ about who they choose to share their lives with.

6. Thanks to our experience and ______________, we have great confidence in our products.

7. Leizu is a ____________ figure in the history of Chinese sericulture.

8. The West Lake first submitted for ___________ on the World Heritage List as "Cultural Properties" in 1999.

9. ___________ is the oldest method of manufacturing yarn.

10. This ancient building still looks as it did ___________.

11. A third way in which a monopoly is created is through the development of patents, which allows a company to ___________ control and use a discovery for seventeen years.

12. The difference between ____________ animals and wild ones lies in whether or not possessing this human quality.

II Render the following Chinese expressions into English. The first letter of each word has already been given.

1. 丝绸工艺 s____________ c____________
2. 家蚕 d____________ s____________
3. 主要生产商 d____________ p____________
4. 生丝 r____________ s____________
5. 无可比拟的影响力 u____________ i____________
6. 非物质文化遗产 I____________ C____________ H____________
7. 被奉为；被尊称为 be r____________ as
8. 组成 c____________ of / be c____________ of

Translation

Translate the following paragraph into English.

据传，嫘祖是中国历史上第一位养蚕和制作丝织品的人，并且她将养蚕技术推广至全国，因而人们尊称她为“蚕女神”。在汉代，“蚕女神”被拟人化，人们将她制成真人大小的雕像以示纪念。皇族们也非常尊敬蚕女神。皇后会帮助采集可制造最好丝绸所需的桑叶，为蚕女神举行纪念仪式，并献上猪和羊等祭品。在大约公元3世纪，建造了一座蚕宫，由皇后监督管理蚕宫的养蚕工作。

Passage 2

Sericulture and the Silkworm

Polly Dornette

A Chinese proverb states, "With time and patience the mulberry leaf becomes a silk gown." A true statement; however, a little help from the silkworm Bombyx mori is also required.

Silkworm Life Cycle

Silkworm eggs are lemon yellow when laid, but turn white or black within a few days. White eggs are infertile and black eggs are **fertile**. Once laid, eggs typically hatch in about 2 weeks, but the **incubation** period is temperature dependent. Freshly hatched **caterpillars** (larvae), also called silkworms, look like small black **strings**.

Once hatched, silkworm larvae normally eat only mulberry leaves; however, they will consume an artificial diet when raised in **captivity**. As they grow and mature, caterpillars become **greyish** white in colour. The caterpillars moult 4 times passing through 5 **larval instars** (an instar is the stage between moults) over a period of 23 to 28 days. During this time, the larvae grow to be about 3 inches in length, 10,000 times their size at hatching. The caterpillar has 6 legs, just as it will when it becomes a **moth** (adult). Each of a larva's body **segments** has a dot, or a **spiracle**, on its side that is used for breathing.

After the 4th moult, the caterpillar begins to **secrete** a single, continuous silk fibre through its **salivary gland**. The strand of silk produced by a silkworm is usually about 1 mile long. The caterpillar moves its head in a constant figure eight pattern to wind the silk fibre around its body to form a dense cocoon. It takes about 3 days for a caterpillar to spin its cocoon. The caterpillar should not be disturbed during this time as a disruption will cause it to restart its cocoon—or could even kill it. If you hold a newly formed cocoon up to a light, you may be able to see the larva moving its head around inside. In its 2" long cocoon, the larva undergoes metamorphosis to a **chrysalis**.

Approximately 3 weeks later, the organism cuts through the cocoon and emerges as a moth. After emerging, moths release a **reddish** brown **fluid** known as **meconium**, a waste product that could not be released while they were in their cocoons. Moths do not eat or drink and survive only long enough to reproduce. Males and females mate with their **posterior** ends stuck together. Females will lay about 500 eggs in less than a week.

For commercial silk production, cocoons must be **intact**. Newly formed cocoons are placed into hot water to kill the chrysalises and **dissolve** the sticky substance that holds the silk threads together. The silk fibre from the cocoons is wound onto spools. Silk thread can be **dyed**, but it is naturally white or gold in colour.

Historical Significance

Despite their small size, silkworms have played a significant role in shaping cultures throughout Europe and Asia. Sericulture, silkworm cultivation for silk production, has been practiced for over 5,000 years in China and in Europe as early as the 6th century. According to Chinese legend, silk was discovered by an ancient Chinese princess when a silk cocoon fell into her tea. When she tried to remove the cocoon, the silk fibres unravelled in her hand.

The Chinese spun up to 48 silk fibres to create silk threads that were then woven into luxurious silk cloth. Silk's commercial value was quickly realized and the secret was closely guarded for nearly 3 millennia. **Smuggling** silkworms, silkworm eggs, or even white mulberry seeds from China was punishable by death.

The Silk Road, an extensive collection of up to 6,000 miles of trade routes between China and the **Mediterranean** Sea, was established and led to the development of civilizations along its routes. In addition to the fabric that was transported via the Silk Road, religion, culture, technology, and even disease were exchanged among civilizations that included **Persia**, Greece, **Syria**, Rome, Armenia, and India.

Constantinople, the **Byzantine** city now known as **Istanbul**, began weaving raw silk purchased from China around the 4th century. In the mid-6th century, 2 Byzantine **monks** were able to smuggle silkworm eggs and mulberry seeds to Constantinople in **hollowed** out **canes**, thus allowing the Byzantines to establish European silk production. Byzantine silk dominated silk production in Europe until the 12th century when the industry began booming in Italy. The flourishing Italian silk industry may have been partially responsible for funding the **Renaissance**[1].

Scientific Contributions

In the mid-20th century, German chemist Adolf Butenandt[2] sought to **isolate pheromones from** silkworms. He eventually was able to isolate **bombykol**, a chemical released by the female silkworm moth to attract mates. The isolation, analysis, and ultimately the **synthesis** of this pheromone has proven beneficial for controlling insect populations.

Considerable research has been done into the science of silk. Silk is a protein consisting primarily of **glycine**, **alanine**, and **serine**. The smoothness of silk is due to these 3 amino acids having the smallest size groups of all amino acids (**-H**, **-CH_3**, and **–CH_2OH**, **respectively**). This amino acid chain forms a **pleated-sheet** structure that **is resistant to** stretching, a well-known property of silk.

Attempts to produce a man-made version of silk have resulted in the development of new materials. **Rayon**, previously referred to as **viscose** silk, is one attempt at a **synthetic** form of silk developed in the early 20th century. Rayon has similar dying properties to natural silk and a high lustre, but it lacks the strength of natural fibres. In 1938 a chemist working for DuPont[3] developed **nylon**, which exhibits many of the same properties as silk at a much lower cost. In addition to being woven to make fabric, nylon has been used for toothbrushes, fishing line, and rope.

Words

fertile [ˈfɜːtaɪl] *adj.* 可繁殖的

incubation [ˌɪŋkjuˈbeɪʃn] *n.* 孵化；孵卵

caterpillar [ˈkætəpɪlə(r)] *n.* 毛虫

string [strɪŋ] *n.* 串；带子；线丝

captivity [kæpˈtɪvəti] *n.* 囚禁；被俘；束缚

greyish [ˈɡreɪɪʃ] *adj.* 浅灰色的；微带灰色的

larval [ˈlɑːvl] *adj.* 幼虫的；幼虫状态的

instar [inˈstɑː] *n.* 中间形态；龄（幼虫两次蜕皮之间的虫期）

moth [mɒθ] *n.* 飞蛾；蛾子

segment [ˈsegmənt] *n.* 部分

spiracle [ˈspaɪrəkl] *n.* 气门

secrete [sɪˈkriːt] *vt.* 分泌

salivary [səˈlaɪvəri] *adj.* 唾液的；分泌唾液的

gland [glænd] *n.* 腺

chrysalis [ˈkrɪsəlɪs] *n.* 蛹

reddish [ˈredɪʃ] *adj.* 淡红色的；微红的

fluid [ˈfluːɪd] *n.* 液体

meconium [mə'kəʊnɪəm] *n.* 蛹便
posterior [pɒ'stɪəriə(r)] *adj.* 后面的；尾部的
intact [ɪn'tækt] *adj.* 完好无损的
dissolve [dɪ'zɒlv] *vt.* 使溶解
dye [daɪ] *vt.* 染色；给……染色
smuggle ['smʌgl] *vt.* 走私；偷运；私运
monk [mʌŋk] *n.* 修道士；僧侣
hollow ['hɒləʊ] *vt. & vi.* 挖空（某物）
cane [keɪn] *n.* 竹杖；藤杖；手杖
Renaissance [rɪ'neɪsns] *n.* 文艺复兴
pheromones ['ferəməʊn] *n.* 信息素；外激素
bombykol ['bɒmbaɪkɒl] *n.* 蚕蛾性诱醇
synthesis ['sɪnθəsɪs] *n.* （物质在动植物体内的）合成
glycine ['glaɪsiːn] *n.* 甘氨酸；氨基乙酸
alanine ['æləniːn] *n.* 丙氨酸
serine ['seriːn] *n.* 丝氨酸
respectively [rɪ'spektɪvli] *adv.* 各自地；各个地；分别地
rayon ['reɪɒn] *n.* 人造丝；人造丝织品
viscose ['vɪskəʊz] *n.* 黏胶
synthetic [sɪn'θetɪk] *adj.* 人造的；（人工）合成的
nylon ['naɪlɒn] *n.* 尼龙

Useful Expressions

isolate sth. from sth. 使（某物质、细胞等）分离；使离析
be resistant to 抵抗的，有抵抗力的；抵制的；阻止的

Proper Names

Mediterranean 地中海的
Persia （地名）波斯（西南亚国家，现在的伊朗）
Syria （地名）叙利亚（西南亚国家）
Byzantine 拜占庭的；拜占庭帝国的

Istanbul （地名）伊斯坦布尔（土耳其西北部港市）

-H 氢原子

$-CH_3$ 甲基

$-CH_2OH$ 甲醇基团

pleated-sheet 折叠片

Notes

1. The Renaissance: the period in Europe during the 14th, 15th and 16th centuries when people became interested in the ideas and culture of ancient Greece and Rome and used these influences in their own art, literature, etc.
   文艺复兴：欧洲14、15和16世纪时，人们对古希腊和古罗马的思想文化产生兴趣，并将这些影响运用到自己的艺术、文学中，以此来繁荣文学艺术。

2. Adolf Butenandt: Adolf Friedrich Johann Butenandt (24 March 1903—18 January 1995) was a German biochemist. He was awarded the Nobel Prize in Chemistry in 1939 for his "Work on Sex Hormones."
   阿道夫・布特南特，全名为阿道夫・弗里德里希・约翰・布特南特（1903年3月24日—1995年1月18日）是一名德国生物化学家，他于1939年凭借"性激素的研究"获得诺贝尔化学奖。

3. Du Pont: E. I. Du Pont de Nemours and Company, commonly referred to as DuPont is an American conglomerate that was founded in July 1802 in Wilmington, Delaware, as a gunpowder mill by French-American chemist and industrialist Éleuthère Irénée du Pont.
   杜邦：杜邦集团，俗称杜邦，是一家美国企业集团，由法裔美国化学家和实业家埃勒泰尔・伊雷内・杜邦于1802年7月在特拉华州威尔明顿成立，其原本是一家火药厂。

Reading Comprehension

I Choose the best answer to each of the following questions or sentences.

1. The incubation period is __________ dependent.
   (A) sunshine (B) air
   (C) moisture (D) temperature

2. After the _______ moult, the caterpillar begins to secrete a single, continuous silk fibre through its salivary gland.
   (A) second (B) third
   (C) fourth (D) fifth

3. The caterpillar moves its head in a constant figure _________ pattern to wind the silk fibre around its body to form a dense cocoon.
   (A) six (B) eight
   (C) two (D) three

4. Silkworm cultivation has been practiced in Europe as early as the___________.
   (A) 5th century (B) 6th century
   (C) 4th century (D) 3rd century

5. ___________ silkworms, silkworm eggs, or even white mulberry seeds from China was punishable by death.
   (A) Smuggling (B) Stealing
   (C) Killing (D) Discarding

6. Which is NOT mentioned as exchanges among civilizations that included Persia, Greece, Syria, Rome, Armenia, and India?
   (A) Disease. (B) Culture.
   (C) Politics. (D) Technology.

7. How did Byzantines establish European silk production?
   (A) They sent someone to China to learn the technique.
   (B) Some travellers brought the silk production back to their country secretly.
   (C) They made independent silkworm research in their own country.
   (D) Two Byzantine monks were able to smuggle silkworm eggs and mulberry seeds to Constantinople in hollowed out canes.

8. Compared with silk, what's the biggest advantage of nylon?
   (A) It is more resistant to stretching.
   (B) It has a higher lustre.
   (C) It owns better dying properties.
   (D) It is at a much lower price.

II Outline the "life cycle of silkworm" in the following chart.

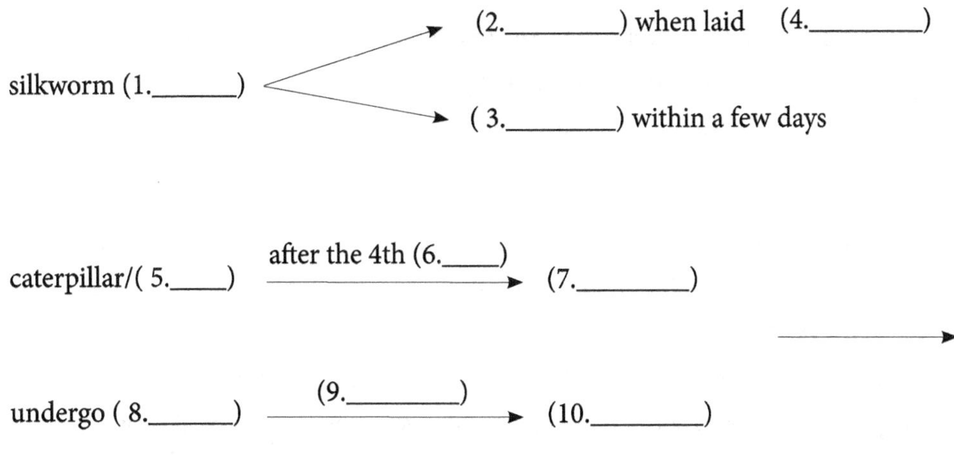

Words and Expressions in Use

I Match the collocations in Column A with the Chinese meanings in Column B.

Column A	Column B
incubation period	A. 大量的研究
spin the cocoon	B. 吐丝结茧
undergo metamorphosis	C. 黏稠的物质
waste product	D. 商业价值
the sticky substance	E. 丝纤维
silk fibre	F. 排泄物
silkworm cultivation	G. 经历变形
commercial value	H. 孵化期
trade route	I. 养蚕
considerable research	J. 商路

II Compare the following pairs of sentences and explain the different parts of speech and meanings of italicized words.

1. lay

(A) She ***lays*** the baby down gently on the bed.

(B) Before they started, they ***laid*** newspaper on the floor.

(C) From a ***lay*** perspective, the article is difficult to get through.

(D) The cuckoo ***lay***s its eggs in other birds' nests.

2. fertile

(A) This leaves ***fertile*** soil unprotected and prone to erosion.

(B) A chess player must have a ***fertile*** imagination and rich sense of fantasy.

(C) Some fish are very ***fertile***, and they lay thousands of eggs.

(D) The region at the time was ***fertile*** ground for revolutionary movements.

3. raw

(A) We import ***raw*** materials and energy and export mainly industrial products.

(B) Generally, ***raw*** vegetables and fruit will contain the most nutrients because cooking can remove some nutrients.

(C) This information is only ***raw*** data and will need further analysis.

(D) He graduated from college only one month ago, so he is a ***raw*** beginner.

(E) This song displays the ***raw*** passions of nationalism.

4. release

(A) Divorce ***releases*** both the husband and wife from all marital obligations to each other.

(B) Police have ***released*** no further details about the accident.

(C) The government has been working to secure the ***release*** of the hostages.

(D) She burst into tears, ***releasing*** all her pent-up emotions.

(E) The **release** of carbon dioxide into the atmosphere is harmful to the environment.

5. property

(A) This campaign fully reveals China's position and determination to protect intellectual ***property***.

(B) A radio signal has both electrical and magnetic ***properties***.

(C) ***Property*** prices are just beginning to harden again.

(D) He was entrusted with the care of her ***property***.

6. consume

(A) The electricity industry ***consumes*** large amounts of fossil fuels.

(B) The hotel was quickly ***consumed*** by fire.

(C) Britons still ***consume*** more than 29 million pints a day, ranging over 1,000 or more different brews.

(D) Carolyn was ***consumed*** with guilt.

III Fill in the blanks with the words or expressions given below. Change the form where necessary. Each word or expression can be used only once.

captive	approximate	emerge	luxurious	punish	seek
respective	smooth	analyse	resistant to	fertile	isolate...from

1. Cotton is more ______________ being squashed and polyester is more resilient.

2. China's production of raw silk accounts for ____________ more than 70% of the total amount in the world.

3. Even though they are bred in the wild, the eggs are hatched in ___________.

4. Although he has ___________ to find a peaceful solution, he is facing pressure to use greater military force.

5. The index in rural areas and third-tier cities edged up by two points and one point ____________. In other areas, it was mainly stable.

6. Jin Li, vice president of Fudan University, said bioscience is now entering an "age of discovery" with the ____________ of new findings.

7. The polluted waste is often dumped, making the surrounding land _______________.

8. These kinds of criminal acts that endanger state security are ______________ in any country.

9. He was immediately _______________ the other prisoners.

10. Now we'll be able to live in____________ for the rest of our lives after winning the lottery.

11. I just love the _____________ of silk.

12. His ____________ really hit home. I had never seen myself in that light before.

Translation Skills

词语英译（2）——根据语境选择译文用词

词语的搭配（collocation）是在英语写作和汉译英时特别要注意的地方。很多时候，表达似乎从语法层面并没有错误，但却是怎么看怎么别扭，原因在于用词不地道，搭配不当。请看以下几个例子。

（1）复习知识。

学生使用频率很高的短语，大部分的译文是review knowledge，但是这却是一个很中式的表达，并不符合英文的搭配习惯，正确的译法是do revisions。

（2）公司提供福利。

此处的“福利”，很多人会译成welfare，但welfare这个词在英文词典中解释为well-being或者是financial or other assistance to an individual or from a city, state, or national government，显然更多的是从社会、政府等层面来理解，所以这个译文并不合适这个短语的语境，较为地道的译文可以是job-related benefits 或者是perks （这个词意为special benefits that are given to people who have a particular job or belong to a particular group）。

（3）废水的排放。

“排放”这个词的一个常见译法为emission，如emission of green house gases，那么“废水的排放”是否就是emission of waste water？我们查阅牛津词典会发现 emission的一个释义为the production and discharge of something, especially gas or radiation，可见此排放主要是针对气体和辐射而言，废水显然不在这个范畴。但在此英文释义中我们可以发现表示排放的另一个词discharge。查询词典会发现discharge在表示排放时，包括了液体的排放（allow a liquid, gas, or other substances）to flow out from where it has been confined），因而discharge of waste water才是合适的译文。

上文已经提到，汉语中同一个词在不同的语境中表达的意思往往有不同的侧重点，因而也会有不同的相对应的地道的英文表达， 请看下面的例子。

（1）先进的学生。

“先进”修饰技术时，常常译为advanced，但在这里是修饰学生，应该是表示进步快、水平高、可以作为学习榜样的，与 advanced表达的前沿的、高水平之意不符，而应该译成high-achieving。

（2）基本工资 vs. 基本利益。

一词多义在汉语中很常见，我们翻译这两个短语时就应该充分考虑这两个“基本”的不同含义，断不能简单地都译成basic，基本工资可以译成basic wage， 基本利益则为fundamental interest。

（3）社会的监督者。

"社会的监督者"很多时候译为"the supervisors of the society",百度百科对"社会监督"所给出的解释为"由国家机关以外的社会组织和公民对各种法律活动的合法性进行的不具有直接法律效力的监督"，而supervisor这个词尤指工人或学生的监督者，所以并不能传达出原文真实的含义。英文当中其实有一个比较对应的词whistle-blower（a person who informs on a person or organization regarded as engaging in an unlawful or immoral activity），选用这个词显然比较地道。另外还需注意的是，如果特别是指媒体的监督者，则需选用另外一个词 gatekeeper。

（4）南京云锦浓缩了中国丝织技艺的精华，是中国古代三大名锦之一。

来看下以下两句译文：

Reflecting the quintessence of Chinese silk weaving skills, Nanjing yunjin, is one of the three well-known brocades made in ancient China.

Encapsulating the quintessence of Chinese silk weaving skills, Nanjing yunjin, is one of the three well-known brocades made in ancient China.

两句对浓缩一词的翻译有不同的选择，浓缩在这里就是集中、概括、凝练的意思，reflect和encapsulate都正确表达了原文的意思，不会引起误解，但相比之下，encapsulate（express the essential features of something succinctly）显然在表达上更为精准、对等。

Exercising Your Skills

I Translate the following phrases into English.

1. 扩大经营

2. 完成订单

3. 失去个性

4. 高产农作物

5. 第三产业

II Tell if the following English translations of the Chinese phrases are acceptable and explain your reasons.

1. 失去自我　lose oneself

2. 学习知识　learn knowledge

3. 文化的传承　inheritance of cultures

4. 调配资源　mobilize resources

5. 接触不同的文化　contact different cultures

Unit Project

The silkworm is the larva or caterpillar or imago of the domestic silk moth. It is an economically important insect, being a primary producer of silk. The business of raising silkworms and unwinding cocoons is now known as silk culture or sericulture.

China is unique in that it created an agricultural industry known as sericulture which included raising silkworms, growing the mulberry leaves needed as feed, processing the cocoons and weaving the silk cloth. Silk played such an important role in Chinese life and history that it came to symbolize "industry" itself.

In addition to clothing, the fabric had other uses. During a time where the barter system was still heavily relied on, silk was used as a currency in both foreign and domestic trade. Government officials were known to receive their salaries in the form of silk bolts, and farmers could pay taxes with the fabric.

Apart from what has been mentioned before, currently, the silkworm is also used in science. It has become a model organism in the study of lepidopteran and arthropod biology. Fundamental findings on pheromones, hormones, brain structures, and physiology have been made with the silkworm. Recently, research is focusing on genetics of silkworms and the possibility of genetic engineering.

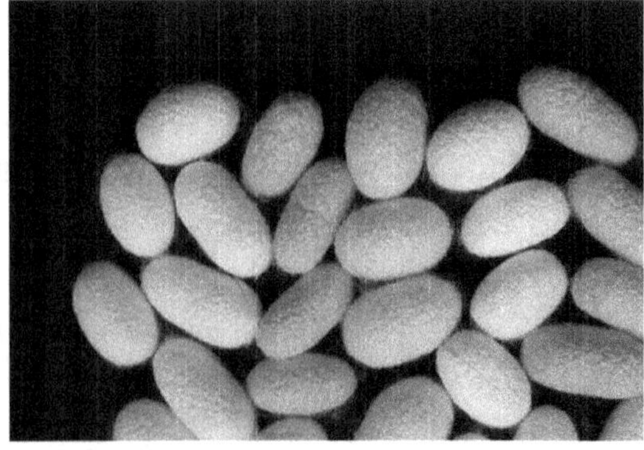

Therefore, silkworm is extremely important in our life and it can be applied into various aspects in the society.

Work in groups to complete the following task:

Work together to search the internet or library for the functions and usages of silkworm and present your report to the class.

Unit 2 Keys

Unit 3 Varieties and Techniques

Unit Guide: Fabrics made of silk consist of many types: brocades, damasks, gauzes, satins, tabbies, etc. due to different weaving techniques. Some are lined, some are unbleached, some are heavy, and some are thin. Whatever way they are made, they are great Chinese contributions to the world culture. In **Unit 3**, we will look at varieties of silk fabric and techniques of silk weaving, and particularly at Kesi, the Chinese silk tapestry and the most common material for the imperial robe. As an ancient craftsmanship, Kesi has its own special loom that cannot be replaced by modern machines and is the best preserved work among all the silk art that has survived to this day.

Lead-in

I Test your knowledge about silk.

Choose the best answer to each of the following sentences.

1. The four prominent embroidery products in China are______. (A) Xiang, Su, Yue and Shu embroidery (B) Xiang, Su, Yun and Shu embroidery (C) Hang, Su, Yue and Shu embroidery (D) Xiang, Su, Yue and Min embroidery	

2. Yunjin Silk Brocade got its name from______. (A) Yun that means Yunnan Province (B) its long history (C) Yun that means cloud (D) its lightness	
3. Shu Brocade is one of the three most famous brocades in China. Shu refers to______. (A) being comfortable (B) being cool (C) Jiangsu Province (D) Sichuan Province	
4. Brocade produced in Suzhou is called______. (A) Song Brocade (B) Su Brocade (C) Jiangsu Brocade (D) Shu Brocade	
5. Ancient weaving techniques roughly divided into two kinds:______. (A) weaving and knitting (B) horizontal weaving and vertical weaving (C) weaving and embroidery (D) single-colored thread and multiple-colored thread weaving	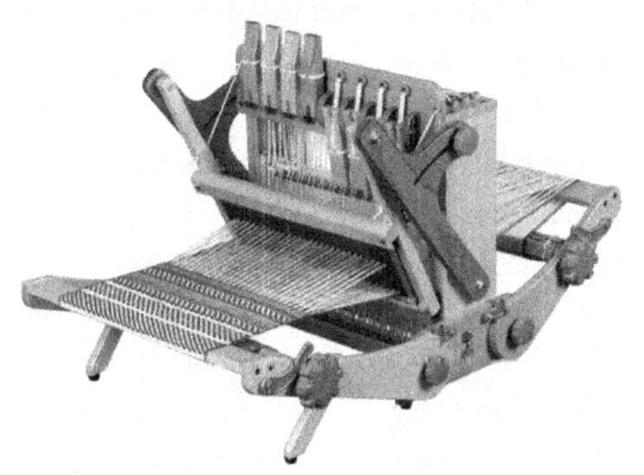

6. The colour of silkworm cocoons depends on ______. (A) what color they are dyed (B) how long they are fed (C) what their caterpillars eat (D) what the silkworms eat	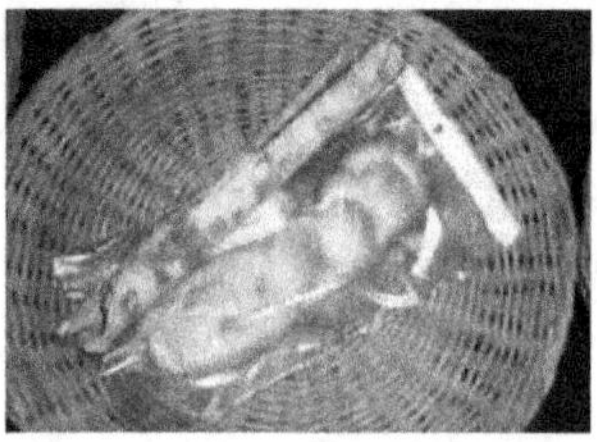
7. The main natural red dye in ancient China was the substance received from_______. (A) minerals (B) animals (C) chemicals (D) plants	
8. The silk dress worn by Indian women is called ______. (A) Silk Saree (B) Silk Qipao (C) Silk Gown (D) Silk Ao Dai	
9. Except as fabrics for garments, the Song brocade can also be used as canvases (帆布) for_______. (A) bags and shoes (B) furniture and decoration (C) paintings and calligraphy works (D) paintings and bags	
10. Because of its softness and natural sheen (光泽), silk has been associated with______throughout history. (A) gentleness (B) luxury and the upper classes (C) color (D) affordability	

Ⅱ A brief introduction to silk production.

Step One: Watch the video and fill in the blanks with appropriate words.

视频1

The floor lamp 1. "________," Fan Weiyan created won him the title "Star of Tomorrow" at the international 2.________ competition earlier this year in Milan.

First appearing in 3.________, the popularity of Kesi tapestry 4.________ during the Ming and Qing dynasties.

It was actually introduced to China from 5.________ along the Silk Road. Then the ancient Chinese added their 6.________ and lifted the technique to a new level.

The biggest harvest Fan got in Hermes is the idea of inheritance and innovation of 7.________. Fan's designs break the traditional form of 8.________ subjects.

Fan is leading his team 9.________ a wider application for the cut silk craft. The designer noted that innovation means better inheritance, and 10.________ the cultural relic with the power of designing is his responsibility.

Step Two: Discuss the following two questions.

1. Why aren't there any styles of Chinese crafts in some world-famous brands like Hermes and others?

2. What personal qualities do we need to do well in Kesi?

Ⅲ Kesi Silk: World Cultural Heritage of Suzhou.

视频2

Watch the video and answer the following questions.

1. Except for the classical gardens, Suzhou is also famous for its silk fabrics. What are the most famous three silk products?

2. What made Kesi be put on the list of World Cultural Heritage?

3. What are involved in Kesi, the ancient silk-weaving technique?

4. Why can Kesi silk fabric widely applied to apparel, paintings, ornaments, and household goods?

5. When can the origin of Kesi silk be traced back to?

Passage 1

Weaving a World of Beauty like Ancients

Qu Zhi

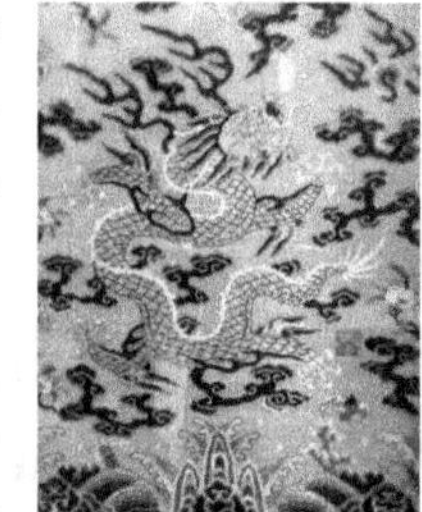

It took five weavers at least three years to complete an **imperial robe** in the Qing Dynasty (1616—1911). The robe has 5-clawed dragons[1] surrounded by blue clouds **interspersed** with bats[2] and 12 symbols[3] showing imperial **authority**, including sun, moon, rock, flames and more.

A single robe required thousands of golden lines and peacock feathers for its **lavish** appearance. **Kesi**—Chinese silk tapestry—was the most common material for the robes. The **intricate** patterns were not **embroidered** onto the cloth but instead woven with colored silk. For each centimetre of material, the Kesi weavers needed to **sew** 100 **horizontal** threads.

"You see the **exquisite** texture of silk tapestry, which is the **instrument** of countless ways of using threads, imagination and **painstaking** effort. Even today, not one step of Kesi can be replaced by mass machinery," Wang Jianjiang[4] tells *Shanghai Daily*.

Wang, 51, is a Kesi master whose ancestors made clothes for the Qing Dynasty royal family, including a birthday **gown** for **Empress Dowager Cixi** (1835—1908).

Now he runs a workshop with eight looms in Suzhou, Jiangsu Province. With the looms **inherited** from his father, Wang and four of his descendants **duplicate** clothing worn by China's ancient royal families, fixing antique Kesi for major museums in China and also creating new works.

With a career in silk tapestry of over 35 years, Wang has answered the same question over and over again—what is this handicraft that he is so **dedicated** to?

Kesi is a form of textile that dates all the way back to the Han Dynasty (206 BC—AD 220). It, along with Hangzhou's silk weaving painting, **Fujian's Yongchun paper weaving painting** and **Sichuan's bamboo screen painting**, are called China's "four big textiles[5]."

Unlike other weaving methods, in which the vertical (warp) and horizontal (**weft**) threads extend back and forth completely across the loom, the secret of Kesi is that it's done on a simple loom using a technique in which warp threads fully extend but the weft ones do not.

This skill is known as "whole warp and broken weft.[6]"

With this principle technique, the formation of the pattern is based solely on changes in weaving the weft threads. Silk tapestry can be appreciated on both sides. And since the various **adjoining** colors in the weft are separate, there will be a slim gap along the edges of the forms, which is why tapestry is also known as "**carved** silk."

For many centuries, the imperial court had a monopoly on kesi weaving. In the Ming (1368—1644) and Qing dynasties, great amounts of royal textiles were Kesi embroideries. Especially in the Ming era, Kesi clothes were **exclusive** to the royal family, removed from public use.

"Hence, when talking about traditional crafts in China, Kesi is hardly known by Chinese people," Wang says. "Even the locals in Suzhou have no idea of it."

The craft of Kesi **prevailed** in the Song Dynasty (960—1279) when it was used to make **replicas** of paintings and **calligraphy** works. During this time, the craft was upgraded from daily **commodity** to work of art, according to Wang.

At his workshop in Suzhou, people can see the works that **resemble** tapestries from the Song Dynasty. Woven against light ochre background, one tapestry on display is the branch of a flowering peach tree, with some blossoms open while a dove is resting on the branch, portraying an **intimate** and quite realistic-looking scene from nature. Traces of ink have been added to the outlines.

"Animal and plant were the most common motifs for Kesi artwork, especially in the Song Dynasty with an **association** of good fortune," Wang says.

Realistic paintings are more commonly used than ink paintings as the basis of Kesi tapestries, because they are easier to weave due to realistic paintings' clearer **strokes** and lines and **delineation** of colors.

"Changing layers of one colour is very hard to weave, and to demonstrate the best effect of stroke needs a lot of experience," Wang says.

In the book ***A Dream of Red Mansions*** by **Cao Xueqin**, one of China's four great classical novels, there is a scene describing the **debut** of **Wang Xifeng**, one of the **principal** characters. Wang is **tactful, worldly** and powerful in the **aristocratic** family illustrated in the book.

The most capable woman in this novel wears a colored Kesi jacket in her debut scene. In the same novel, red Kesi textile is given to **Jiamu** (Mother Jia), the most respected character, as a present.

Novelist Cao's great-grandfather and uncle both worked for the textile administrations in Suzhou, so the writer knew the value of Kesi—the **ultimate** luxury of imperial life and aristocrats.

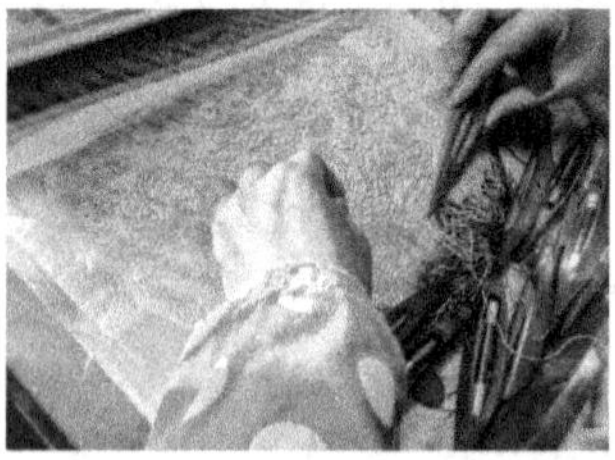

Old poetry **opined** that a woman could wear an outfit in Kesi when she arrived at the end of her life, a hint about how long the process of making it took. And there was an old saying, "An inch of Kesi is worth an inch of gold."

The manufacturing process is extremely complex. First, all the warp threads must be fixed to the tapestry loom. Then the pattern from a painting, for example, is placed underneath the flat and even warp threads, and the weavers will use a brush to outline the forms onto the warp.

After that, a **bewildering** array of coloured silks are prepared according to the **hues** of the **prototype** that the craftsman needs to separate the color and install them in the shuttle **groove**. Using dozens of weaving techniques, the weavers then move the shuttle back and forth between the warp threads.

"Sometimes the work is very repetitive and dull but requires the craftsman to be very patient and **meticulous**," Wang says. "It is indeed an **arduous** work process that if one has no passion for, they won't persist," Wang says.

There is also no room for error. Because it is woven, it cannot be **undone**, and a mistake may mean the weavers must start again **from scratch**.

"It usually takes three years to learn the basic skills of Kesi, while making a good work usually takes over 20 years of **apprenticeship** and practice," says Wang.

Wang's father, Wang Jialiang, is 76 and still making silk tapestry. "I think my father still does a better job than me, since Kesi requires a lot of experience," Wang says.

He approaches one of the works in his workshop and points at one special Kesi tapestry. It has two base colours—gold on one side and silver on the other. One side has the character for **longevity** while there is a peach on the other side. One side was woven with silk and the other with gold thread.

"But they (the two sides) are woven on one layer," Wang says.

Created by Wang's father, the family business is the only one among **a handful of** workshops making Kesi tapestry in China that is able to weave different patterns on the two sides.

"It's a secret that I cannot reveal," Wang says with a smile. "The only one I will teach this skill to is my daughter, who **is bound to** inherit the family **legacy** even if she doesn't like it. But luckily she enjoys it."

Words

imperial [ɪm'pɪərɪəl] *adj.* 帝国的；皇帝的

intersperse [ɪntə'spɜːs] *vi.* 点缀；散布

authority [ɔː'θɒrɪtɪ] *n.* 权威；权力

lavish ['lævɪʃ] *adj.* 丰富的；大方的

intricate ['ɪntrɪkət] *adj.* 复杂的；错综的

embroider [ɪm'brɒɪdə] *vt.* 刺绣；装饰

sew [səʊ] *vt.* 缝合，缝上

horizontal [hɒrɪ'zɒntl] *adj.* 水平的；地平线的

exquisite ['ekskwɪzɪt] *adj.* 精致的；细腻的

instrument ['ɪnstrʊmənt] *n.* 工具；手段

painstaking ['peɪnzteɪkɪŋ] *adj.* 艰苦的；勤勉的

gown [gaʊn] *n.* 长袍；礼服

dowager ['daʊədʒə] *n.* 贵妇；继承亡夫爵位的遗孀

inherit [ɪn'herɪt] *vt.* 继承

duplicate ['djuːplɪkeɪt] *vt.* 复制

dedicate ['dedɪkeɪt] *vt.* 致力；献身

weft [weft] *n.* 纬纱

adjoin [ə'dʒɒɪn] *vi.* 毗连；邻接

carve[kɑːv] *vt.* 雕刻

exclusive [ɪk'skluːsɪv] *adj.* 独有的；专一的

prevail [prɪˈveɪl] *vi.* 盛行；流行
replica [ˈreplɪkə] *n.* 复制品；复制物
calligraphy [kəˈlɪgrəfɪ] *n.* 书法；笔迹
commodity [kəˈmɒdɪti] *n.* 商品；日用品
resemble [rɪˈzembl] *vt.* 类似；像
intimate [ˈɪntɪmət] *adj.* 宁静怡人的；亲密的
association [əˌsəʊʃɪˈeʃən] *n.* 联想
stroke [strəʊk] *n.* 笔画
delineation [dɪˌlɪnɪˈeɪʃən] *n.* 描述；画轮廓
debut [ˈdeɪbjuː] *n.* 初次登台
principal [ˈprɪnsəpəl] *adj.* 主要的
tactful [ˈtæktfʊl] *adj.* 机智的；圆滑的
worldly [ˈwɜːldli] *adj.* 世俗的；尘世的
aristocratic [ˌærɪstəˈkrætɪk] *adj.* 贵族的
ultimate [ˈʌltɪmət] *adj.* 最终的；极限的
opine[əʊˈpaɪn] *vt.* 以为；想
bewilder [bɪˈwɪldə] *vt.* 使迷惑；使不知所措
hue[hjuː] *n.* 色调；色彩
prototype [ˈprəʊtətaɪp] *n.* 原型；标准
groove [gruːv] *n.* 凹槽
meticulous [məˈtɪkjʊələs] *adj.* 一丝不苟的；小心翼翼的
arduous [ˈɑːdjʊəs] *adj.* 努力的；费力的
undo [ʌnˈduː] *vt.* 取消；破坏
apprenticeship [əˈprentɪsʃɪp] *n.* 学徒期；学徒身份
longevity [lɒnˈdʒevɪti] *n.* 长寿，长命；寿命
handful [ˈhændfʊl] *n.* 少数；一把
legacy [ˈlegəsi] *n.* 遗赠；遗产

Useful Expressions

imperial robe 龙袍
from scratch 白手起家；从头做起
be dedicated to 奉献；从事于，致力于
a handful of 极少数；少量

be bound to　必然；一定要

Proper Names

Kesi　缂丝，又称“刻丝”“克丝”，中国最传统的一种采用“通经断纬”织法的丝织品
Empress Dowager Cixi　慈禧太后
Fujian’s Yongchun paper weaving painting　福建永春纸织画，福建省永春县特产
Sichuan’s bamboo screen painting　四川竹帘画是中国传统手工艺品，曾被列为皇家贡品
A Dream of Red Mansions　《红楼梦》，中国古代章回体长篇小说
Cao Xueqin　曹雪芹
Wang Xifeng　王熙凤
Jiamu (Mother Jia)　贾母

Notes

1. Chinese dragons are legendary creatures in Chinese mythology and folklore（民间传说）. In the historic magnitude of Chinese art, dragons are typically portrayed as long, scaled, serpentine（蜿蜒的；蛇形的）creatures with four legs. In the Qing Dynasty, the 5-clawed dragon was assigned to represent the Emperor while the 4-clawed and 3-clawed dragons were assigned to the commoners（下臣）.
民间“五爪为龙，四爪为蟒”的说法形成于清代，主要作为皇帝与下臣服装上纹饰的差别，皇帝穿“龙袍”，其他皇族和下臣穿“蟒袍”，但这只是名称上的差别而已，从龙的形式上来讲无论龙和蟒都是四足蛇类，形状无差异。

2. Bats have the symbolic meaning of happiness, longevity and peace in Chinese traditional culture.
在中国的传统文化中，蝙蝠是幸福、长寿与和平的象征，蝙蝠又称“遍福”。

3. “Twelve symbols” are twelve kinds of decorative patterns on ancient emperors’ robes and auspicious clothes, which are sun, moon, stars, mountain, dragon, golden pheasant, fu (square patch on official costume embroidered with white and black axes), fu (an embroidery in square pattern on official gowns), zongyi (two goblets with the image of tiger and long-tailed ape), seaweed, fire, and rice.
“十二章”是古代帝王礼服和吉服上的十二种装饰纹样，依次是日、月、星辰、群山、龙、华虫、黼、黻、宗彝、藻、火、粉米。

4. Three generations of the Wang family, living in Suzhou, were the kesi fabric craftsmen serving the Qing Government. The sixth generation of successor Wang Jianjiang has once repaired the kesi fabric artifacts in the Forbidden City and copied the dragon robes worn by the emperors of the Yuan, Ming and Qing dynasties for the Capital Museum.
王建江，苏州王氏缂丝世家第六代传人，非物质文化遗产传承人，高级工艺美术师，多次参与故宫博物院缂丝文物的修复与复制。

5. Four big textiles are the four major traditional Chinese textiles in history, including Suzhou's Kesi painting, Hangzhou's silk weaving painting, Fujian's Yongchun paper weaving painting and Sichuan's bamboo screen painting.
历史上，苏州缂丝画、杭州丝织画、福建永春纸织画和四川竹帘画并称为中国的"四大家织"。

6. With a combination of partial weft threads attached to full warp threads, Kesi differs from other woven silk goods that have full warp and weft threads. In other words, the warp threads are first set in place and then threads of various colors and lengths are shuttled and attached for the weft to create the desired design or pattern. When seen up close, the colored threads that make up the images are all independent. Therefore, it is particularly difficult to make patterns with curves using only vertical weft threads. Furthermore, since the edges of forms and patterns often appear abrupt and have a saw-tooth gap where they are attached to the warp, they look as if they are cut out. For this reason, this fine silk tapestry is also referred to as "cut silk" in Chinese.
缂丝是一种用"通经断纬"方法织造的丝织品，不同于一般锦的织法"通经通纬"法，即纬线穿通织物的整个幅面。所谓"通经"，就是用本色丝线做经线；所谓"断纬"，就是用各种彩色丝线做纬线。经纬线相交，根据纹样的图案变化。缂织时，先在织机上安装好经线，经线下衬书稿或画稿，使用多种色彩的纬线自由转换，制作出画样的彩色图案。由于彩纬充分覆盖于织物上部，织后不会因纬线收缩而影响画面花纹效果。其次，经过彩纬显现花纹，形成花纹边界，呈现出犹如雕琢镂刻的效果。织物上花纹与素地、色与色之间呈现一些断痕，类似刀刻的形象，因此，缂丝也称为"刻丝"。

Reading Comprehension

Choose the best answer to each of the following questions or sertences.

1. Kesi cannot be replaced by mass machinery, because _____.
(A) the golden lines and peacock feathers cannot be used on machines

(B) the complex patterns should be embroidered onto the cloth

(C) numerous ways of using threads are applied in the exquisite texture of this silk tapestry

(D) it's too expensive for mass production

2. Different from other weaving methods, the secret of Kesi is that _____.

(A) it needs a set of looms

(B) the vertical threads fully extend but the horizontal ones do not

(C) the horizontal threads fully extend but the vertical ones do not

(D) the warp and weft threads extend back and forth completely across the loom

3. Which of the following statements is NOT true?

(A) The formation of the Kesi pattern is based both on changes in weaving the weft and warp threads.

(B) The pattern of Kesi can be viewed and enjoyed by the viewers on both sides.

(C) Kesi is also known as "carved silk."

(D) Kesi was too expensive for ordinary people in history.

4. In the ___ Dynasty, wearing Kesi clothes is the privilege only for the royal family.

(A) Qing (B) Ming

(C) Han (D) Song

5. According to Wang Jianjiang, the Kesi craft turned into work of art from daily commodity due to ___.

(A) the spread of the craft in the Song Dynasty

(B) the wearing of Kesi among ordinary people

(C) the teaching of the craft to normal people

(D) the making of imperial robes

6. ______ are commonly used for the major themes of Kesi artwork.

(A) Sun and moon (B) Dragons

(C) Animal and plant (D) Bats

7. A woman could wear an outfit in Kesi when she arrived at the end of her life, which possibly implies that______.

(A) only old women could wear Kesi

(B) only old women could make Kesi because they are more skilled

(C) Kesi only appeared in imperial life and aristocrats

(D) it took the craftsmen a long time to make Kesi

8. According to Wang Jianjiang, the only person who can inherit the Kesi skill is ____.
   (A) his son (B) his apprentice
   (C) his daughter (D) his relative

Summary

The following is a summary of Passage 1. Fill in the blanks with appropriate words.

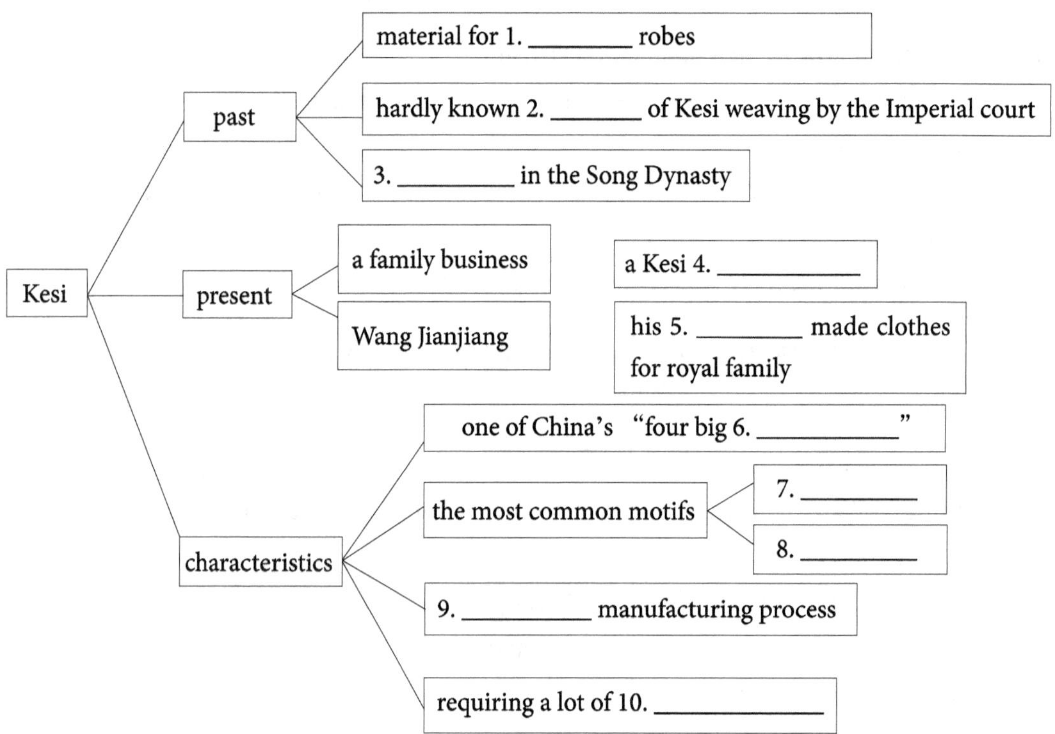

Word Building

I The prefix *re* can be used together with other words to form new ones, meaning "back, intensive, again, against, away, mutually." Match the following words with their parts of speech and Chinese meanings.

	replica	决心
	repetitive	无情的；不间断的
	reminisce	善于接受的
n.	relentless	复制；再生
	resolution	回忆
adj.	repress	复制品
	reproduce	后退；撤回
v.	recede	回答；做出反应
	respond	抑制；镇压
	receptive	重复的

II The prefix *inter* can be used together with other words to form new words, meaning "between, among, mutual." Match the following words with their parts of speech and Chinese meanings.

	interrupt	城市间的
	intercept	会见；面试
	intersperse	干预
n.	interrogate	点缀；散布
	intersect	国际的
adj.	intercity	拦截
	interfere	横断
v.	interchange	审问；质问
	international	交换；互换
	interview	打断；打扰

Words and Expressions in Use

I Fill in the blanks with the words or expressions given below. Change the form where necessary. Each word or expression can be used only once.

narrate	give rise to	manufacture	original	textile	traditional
intangible	add	sew	weave	unique	development

1. The collection mixes traditional Chinese ________ techniques with laser-cut technology, handmade silk and six different types of leather.

2. After investigation, traditional craft projects with profound cultures, excellent ________ processes and sound market prospects in Shandong Province have been found.

3. They will also invite local craftspeople (手艺人) to perform _____ cultural heritage in the centre, ranging from woodcarving and weaving, to making traditional oil-paper umbrellas and creating silver jewellery.

4. The spinning and weaving workshop on Cultural Street, where elderly local ladies can be seen working every day, attracts visitors to watch them practice a vast array of ______ arts unique to the Jiangnan region.

5. Grandma Tea is a folk custom specific to Zhouzhuang where local women gather together in their homes or in tea houses to drink tea, eat, chat, ________ and knit.

6. Exploring the relationship between the organic and the inorganic, and developments in biology and technology, Van Herpen combines __________ weaving techniques with cutting-edge digitally designed weaving on coats and dresses.

7. Hand-made embroidery is known as slow art, in which artists use their hands to weave items for everyday use, ________ history and traditions, or explore the past, present and future.

8. Influenced by national folk art, Yue embroidery formed its own _________ characteristics.

9. From the magnificent Dragon Robe worn by Emperors to the popular embroidery seen in today's fashions, embroidery _________ so much pleasure to our life and our culture.

10. Embroidery is, in its long ________ , inseparable from silkworm-raising and silk-reeling and weaving.

11. The production of silk thread and fabrics _______ the art of embroidery.

12. Some of the costumes and props (道具) used in the drama are delicate reproductions of the ________ ones in Qing Dynasty.

II Render the following Chinese expressions into English. The first letter of each word is given.

1. 复杂的图案 i____________ p____________
2. 精致的纹理 e____________ t____________
3. 机器化大生产 m____________ m____________
4. 四大家织 f____________ b____________ t____________
5. 通经断纬 w____________ w____________ and b____________ w____________
6. 刻丝 c____________ s____________
7. 日用品 d____________ c____________
8. 好运 g____________ f____________
9. 写实画 r____________ p____________
10. 从头开始 s____________ f____________ s____________

Translation

Translate the following paragraph into English.

缂丝是中国传统丝绸艺术品中的精华。自宋元以来，皇室垄断了它的生产。缂丝常用于帝后服饰、皇室肖像和知名画作的织造。因其织造过程极其细致，品格富贵高雅，缂丝素有“一寸缂丝一寸金”“织中之圣”之称。缂丝采用“通经断纬”的织法，使它具有如同雕琢镂刻的效果，且富双面立体感。缂丝工艺极为精湛，在唐代已登峰造极，工匠们已能使用纯金线、纯银线、孔雀羽毛等多种名贵的材质进行交汇缂织。

Passage 2

Classification of China Silk Textiles

The China silk textiles[1] are generally known as ling (twill damask), luo (**gauze**), juan (silk **tabby**) and duan (**satin**). They are further subdivided into juan (silk tabby), qi (damask on tabby), luo (gauze), sha (plain gauze), ling (twill damask), jin (**polychrome** woven silk) and Kesi (silk tapestry with cut designs). Today, according to the silk material, texture, weave structure and texture, the silk textiles can be divided into ling (twill damask), luo (gauze), juan (silk tabby), zhou (**crepe**), jin (brocade), duan (satin), ni (**matelassé**), fang (**habotai**), xiao (**chiffon**), ge (**bengaline**), rong (**velvet**), sha (plain gauze) and ti (**crepons**).

Ling (twill damask)

The basic character of ling is the **diagonal** lines on China silk textiles. The diagonal lines have two types: simple twill and twill damask. The simple twill is the basic or irregular twill, and the twill damask refers to the use of twill weave as the silk ground. It was very popular in Tang Dynasty (618-907), among which the best-known type was Liaoling damask.

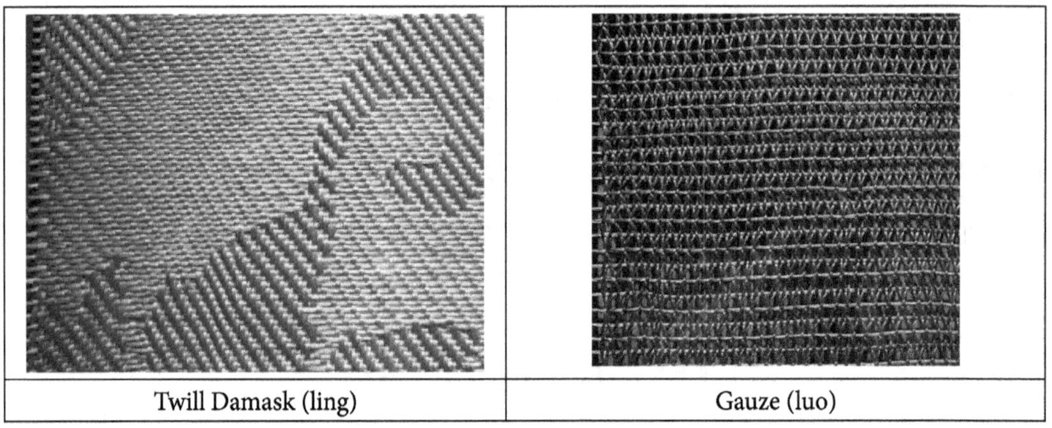

Twill Damask (ling)	Gauze (luo)

Luo (gauze)[2]

Luo **refers to** the textiles woven with stranded warps. It appeared in the Shang Dynasty (16th century BC—11th century BC). In Tang Dynasty, the Yueluo gauze, produced in Yue State in today's Zhejiang Province, and single-warp gauze are quite famous.

Juan (silk tabby)

In ancient times, juan is the general name for the light plain weave fabrics with close texture. It **was originated in Neolithic period** (6000 BC—2000 BC) and is still practiced today.

Duan (satin)

Duan refers to the silk fabrics where only one silk thread (weft or warp) floats on the surface, making the surface of the cloth extremely bright and smooth. It was first seen in the Yuan Dynasty (1206—1368) and became chief product in Ming and Qing dynasties.

Silk Tabby (juan)	Satin (duan)

Famous China Silk Products

Yun Brocade[3]

Yun Brocade was developed during the Yuan Dynasty although its origin could **date back to** the Southern Dynasty. Historical records suggest that **prior to** the Southern Dynasty there was no brocade in Nanjing until **Emperor Liu Yu** (363—422) had workers move to **Jiankang** (today's Nanjing) and established production there. Yun Brocade features quality material, refined weaving and the wide use of gold and silver threads. This **magnificence** gained the name of Yun, which is Chinese for "cloud. " Elegant Yun Brocade **ultimately** became a precious artwork and no longer a fabric for day to day wear. In the Yuan, Ming and Qing dynasties Yun Brocade was a royal **tribute**.

Shu Brocade

Shu Brocade made in Sichuan **originated from** Han and reached its **heyday** in the Wei, Jin, Sui and Tang dynasties. With red as the **predominating** colour, Shu Brocade has a variety of designs, fully reflecting the flowery nature of Shu Culture. In the Tang Dynasty, Dou Shilun[4],

Duke Lingyang, created a set of designs for Shu Brocade, which was known as the "Duke Ling Yang Pattern. "

As the unique weaving skill cannot be **undertaken** by modern machinery, Yun Brocade remains very expensive.

Song Brocade

Song Brocade, as the name implies, originated from the end of Northern Song. Record has it that a Mr. Ji used Song Brocade for the **paperhanging** of some precious calligraphy copies. In all there were twenty different designs of this brocade. When the Ji family declined, they wanted to sell the copies but failed to do so **due to** the high price they demanded. Later some rich man bought them and used the brocade as patterns in workshops in Wu (today's Suzhou area). He made a substantial profit from it! This is regarded as the origin of Song Brocade. And it inherits the old tradition—today Song Jin is mainly used for paperhanging.

Brocades by Ethnic Groups[5]

Zhuang Brocade

Zhuang Brocade is the creation of the Zhuang ethnic group in Guangxi. Zhuang Brocade features rich colours, as this is the **artistry** of Zhuang people. The patterns vary from waves, clouds, grass and flowers to animals. Phoenix, the symbol of **auspiciousness**, takes a dominant role in these designs. Zhuang Brocade is **durable** and can be widely used in **bedding**, belts, bags and clothes.

Yao Brocade from Hunan is **notable**, too. The Brocades for bedding are neat and light in colour with simple designs while those for clothes are flowery. Some of the symbols woven into Yao Brocade are regarded as the legendary "female characters, " which are limited among women. Women use these special symbols or characters to convey messages. Though there is no way of telling if the female character is a branch of an ancient **oracle** system or that of Yao words, Yao Brocade has become a medium of such a mystery.

Hang Brocade made in Hangzhou, Jing Brocade from Jiangling, Hubei, and Ning Silk from Nanjing, Jiang Silk form Zhenjiang, Jiangsu Province, Hu Crape from Huzhou, Zhejiang Province all occupy outstanding roles in the history of Chinese silk.

Words

gauze [gɔːz] *n.* 纱布；薄纱
tabby [ˈtæbi] *n.* 平纹
satin [ˈsætɪn] *n.* 缎子；缎子衣服
polychrome [ˈpɒlɪkrəʊm] *adj.* 多彩的；多色装饰的
crepe [kreɪp] *n.* 绉纱；绉绸 (=crape)
matelassé [mætˈlæseɪ] *n.* 麦特拉斯提花花式织物；呢
habotai [ˈhabətaɪ] *n.* 电力纺
chiffon [ˈʃɪfɒn] *n.* 雪纺绸
bengaline [ˈbeŋgəliːn] *n.* 罗缎；孟加拉织品
velvet [ˈvɛlvɪt] *n.* 天鹅绒；丝绒
crepon [ˈkrepɔn] *n.* 重绉纹之织物；厚绉纱
diagonal [daɪˈægənəl] *adj.* 斜的；斜纹的
magnificence [mægˈnɪfɪsns] *n.* 壮丽；华丽
ultimately [ˈʌltɪmətli] *adv.* 最后；基本上
tribute [ˈtrɪbjuːt] *n.* 礼物；[税收] 贡物
heyday [ˈheɪdeɪ] *n.* 全盛期
predominate [prɪˈdɒmɪneɪt] *vi.* 占主导（或支配）地位；占优势
undertake [ʌndəˈteɪk] *vt.* 承担；从事
paperhanging [ˈpeɪpəhæŋɪŋ] *n.* 裱糊
artistry [ˈɑːtɪstri] *n.* 艺术性；艺术技巧
auspiciousness [ɔːˈspɪʃəsnis] *n.* 吉兆；幸运
durable [ˈdjʊərəbl] *adj.* 耐用的；持久的
bedding [ˈbedɪŋ] *n.* 寝具
notable [ˈnəʊtəbl] *adj.* 显著的；著名的
oracle [ˈɒrəkəl] *n.* 神谕；预言

Useful Expressions

refer to 指的是
be originated in 起源于
date back to 追溯到
prior to 在……之前；居先

originate from 起源于

due to 由于；应归于

Proper Names

Neolithic period [古] 新石器时代

Emperor Liu Yu 皇帝刘昱（463年3月1日—477年8月1日）

Jiankang 建康（南京在六朝时期的名称）

Notes

1. The China silk textiles are various in types with a great variety of patterns and elegant designs. In addition, the patterns are painstakingly organized to show rich variations and great diversity of forms. These textiles represent the elegance and exquisiteness of traditional Chinese silk textiles.
中国丝绸纺织品品种繁多，绫罗绸（绢）缎是日常生活中对丝织品的通称，并非一个完整的分类方法。中国古代丝织品种有绢、纱、绮、绫、罗、锦、缎、缂丝等。今天，丝织品则依据组织结构、原料、工艺、外观及用途分成纱、罗、绫、绢、纺、绡、绉、锦、缎、绨、葛、呢、绒、绸十四大类。

2. Ling and Luo are generally known as the two major China silk textiles, which are often used in day to day garments and clothes.
绫和罗被称为中国丝绸的两大主要丝织品，常用于日常服装。

3. In ancient China, brocades were the representatives of top-class textiles. The Nanjing Yunjin Brocade has also been the best among the four kinds of the most wonderful brocade (Yunjin, Songjin, Shujin and Zhuangjin). It enjoys the fame as "a unique talent of China" and "rarity of the world." About the name "Yunjin Brocade," "Yun" means "cloud" in Chinese, "Jin" means "brocade." People gave the name "Yunjin" to it since it was as gorgeous and florid as cloud.

在中国古代，织锦是高档丝制品的代表。南京云锦在云锦、宋锦、蜀锦、壮锦四大名锦中也名列前茅，被誉为"中国奇锦""世界奇锦"。云锦因其色泽光丽灿烂，美如天上云霞而得名。

4. Dou Shilun is also known as Duke Lingyang, an artist in Tang Dynasty. The Tang Dynasty witnessed the diversification of motifs and styles of silk patterns. The practice of

presenting western animal patterns in traditional Chinese floral roundels started in early Tang Dynasty, and was named Pattern of Duke Lingyang.
窦师纶，唐代丝织工艺家、画家，字希言，京兆（今陕西西安）人。初为秦王李世民府咨议，相国录事参军，后官至大府卿，封陵阳公。他曾研究过舆服制度，精通织物图案设计，被唐政府派往盛产丝绸的益州（今四川省）大行台检校修造。他在继承优秀传统图案的基础上，吸收中亚、西亚等地的题材和表现技法，洋为中用，创造出寓意祥瑞、章彩奇丽的各式新颖绫锦，在当时极为流行，被誉为“陵阳公样”。

5. Brocades by ethnic groups are also the precious Chinese intangible cultural heritage. In China, there are eight most famous types of brocade, namely, Miao brocade, Zhuang brocade, Yao brocade, Dong brocade, Dai brocade, Buyi brocade, Tujia brocade and Maonan brocade.
现有的中国少数民族织锦主要有8种，分别是苗锦、壮锦、瑶锦、侗锦、傣锦、布依锦、土家锦和毛南锦。其中以壮锦最为有名，与云锦、蜀锦、宋锦并称中国四大名锦，是中华民族文化瑰宝。

Reading Comprehension

I Answer the following questions according to Passage 2.

1. What are generally known types of the China silk textiles?

2. What is the basic character of ling?

3. In which dynasty did luo appear?

4. When did juan originate from?

5. When was duan first seen in history?

6. Why did Yun Brocade get its name as we know today?

7. Why does Yun Brocade remain very expensive?

8. What are the features for Zhuang brocade?

9. For what reason is Yao brocade famous and special?

10. According to the passage, what are the most famous four Chinese silk products?

II Outline the "Major Classification of China Silk Textiles" according to different categories.

1. According to the well-known classification and the silk material, there are generally four kinds of silk textiles:
________, ________, ________ and ________.

2. According to the place of origin, there are mainly four famous China silk products:
________, ________, ________ and ________.

3. Brocades by ethnic groups are mainly including ________ and ________ mentioned in the passage.

Words and Expressions in Use

I Match the collocations in Column A with the Chinese meanings in Column B.

Column A	Column B
1. twill damask	A. 斜条纹
2. diagonal lines	B. 书法仿品
3. silk textiles	C. 皇家贡品
4. stranded warps	D. 斜纹花缎
5. close texture	E. 绞股经纱
6. royal tribute	F. 陵阳公样
7. Pattern of Duke Lingyang	G. 质地紧密
8. calligraphy copies	H. 丝织品

II Compare the following pairs of sentences and explain the different parts of speech and meaning of italicized words.

1. warp

(A) The door, ***warped*** by seasons and sea-changes, split slightly.

(B) I never had any toys, because my father thought that they would ***warp*** my personal values.

(C) When a divorced woman re-enters the world of dating and romance, she's likely to feel as though she has entered a time ***warp***.

(D) Another method for making floor coverings involves tying pieces of yarn (纱线) onto the ***warp***.

2. fabric
   (A) Whatever your colour scheme (色调搭配), there's a ***fabric*** to match.
   (B) The ***fabric*** of society has been deeply damaged by the previous regime (政权).
   (C) And there is a necktie, comparable ***fabric***, comparable pattern.
   (D) Condensation (冷冻凝结) will eventually cause the ***fabric*** of the building to rot away.

3. texture
   (A) In the 18th and 19th centuries, the Industrial Revolution transformed the socioeconomic ***texture*** of Britain.
   (B) Earthworms consume large amounts of soil, and produce a rich humus (腐殖质), perfect in ***texture***.
   (C) The very ***texture*** of his prose bears the influence of his familiarity with drugs.
   (D) It is used in moisturizers to give them a wonderfully silky ***texture***.

4. strand
   (A) She tried to blow a gray ***strand*** of hair from her eyes.
   (B) He's trying to bring together various ***strands*** of radical philosophic thought.
   (C) The airport had to be closed, ***stranding*** tourists.
   (D) A ***stranded*** whale with plastic in his belly is seen in Wakatobi, Southeast Sulawesi, Indonesia, in November.

5. thread
   (A) This time I will do it properly with a needle and ***thread***.
   (B) The ***thread*** running through many of these proposals was the theme of individual power and opportunity.
   (C) A thin, glistening ***thread*** of moisture ran along the rough concrete sill.
   (D) Slowly she ***threaded*** her way back through the moving mass of bodies.

6. feature
   (A) The spacious gardens are a special ***feature*** of this property.
   (B) His ***features*** seemed to change, so I could barely recognize him.
   (C) It's a great movie and it ***features*** a Spanish actor who is going to be a world star within a year.

(D) Jon ***featured*** in one of the show's most thrilling episodes.

Ⅲ Fill in the blanks with the words or expressions given below. Change the form where necessary. Each word or expression can be used only once.

heyday	origin	notable	predominate	prior to	durable
date back to	undertake	magnificence	refer to	due to	originate from

1. In his speech, he ________ a recent trip to Canada.

2. They are forced to return to their country of ________.

3. All carbohydrates（碳水化合物）________ plants.

4. The Royal Palace, which ________ the 16th century, is undergoing extensive restoration（复原）.

5. A man seen hanging around the area ________ the shooting could have been involved.

6. The country's economic problems are largely ________ the weakness of the recovery.

7. The proposed new structure is ________ not only for its height, but for its shape.

8. Fine bone china is eminently（显著地）practical, since it is strong and ________.

9. I shall never forget the ________ of the Swiss mountains and the beauty of the lakes.

10. She ________ the arduous task of monitoring the elections.

11. In older age groups women ________ because men tend to die younger.

12. In its ________, the studio's boast was that it had more stars than there are in heaven.

Translation Skills

词语英译（3）—— 词义空缺

词义空缺是指英汉语中相互缺乏对应的词语，由于中西方在文化的各个层面差异较大，很多中国特色的词在英文中很难找到对等的或类似的词，主要集中于名词，这主要是因为文化现象首先是作为一种观念而进入到语言系统中。

处理词义空缺这一现象，常见的翻译策略主要包括以下几种。

1. 音译

在英语文化中完全空缺的事物常会通过音译移植到英语文化中，很多也已经成为了英语中广为接受的表达。例如：功夫kung fu、太极tai chi、豆腐tofu、饺子jiaozi、乌龙茶oolong、乒乓ping pong等。

有时还会在英译的基础上加上注解，例如：词Ci (poetry written to certain tunes with strict tonal patterns and rhyme schemes and in fixed numbers of lines and words, originating in the Tang Dynasty and fully developed in the Song Dynasty)。

2. 直译

和音译相比，直译是解决词义空缺更为稳妥的方法。例如龙舟dragon boat、春节Spring Festival、纸老虎paper tiger、希望工程Hope Project等。

对于有些词，光是直译并不能传递原文的含义，就需要增补缺失的信息，简单的如“保护伞”（在反腐语境中，特指黑势力的保护伞），直译成protective umbrella还不够，而应译成the protective umbrella of mafia-like gangs and evil forces。文化内涵更为丰富的词还需要加注，以补充必要的文化背景知识。例如，虎头蛇尾the head of a tiger and the tail of a snake (a brave beginning but a weak end)、八股文eight-part essay (a literary composition prescribed for the imperial civil service examination, known for its rigidity of form and poverty of ideas)、草根工业grass root industry (refers to village and township enterprises which take root among farmers and grow like wild grass)。

3. 意译

当上述两种策略都不适用时，常会采用意译，虽然无法保留汉语原汁原味的感觉，但胜在能够准确地传递原词的文化信息。例如：拜堂perform the marriage ceremony；形象工程vanity projects；拳头产品knock-out products；盲流jobless migrants from rural areas to cities；不忘初心，方得始终Never forget why you started, and your mission can be accomplished；等等。

意译和加注两种策略也会同时采用，为英语读者提供更为详尽的词语文化内涵。例如：

（1）元宵节the Lantern Festival (the 15th of the 1st lunar month)

（2）如果一只喜鹊停在梅花的枝头，寓意着“喜上眉梢”。

If a magpie (representing good news) rests on the branch of the plum blossom, it has the meaning “happiness up to one’s eyebrows.” (Plum blossom has a similar pronunciation to “eyebrow” in Chinese.)

Exercising Your Skills

Translate the following words or expressions into English using the translation techniques you have just learned.

1. 云锦

2. 压岁钱

3. 七夕节

4. 豆腐渣工程

5. 共享充电宝

6. 狗咬吕洞宾

7. 一带一路

8. 四大名绣

Unit Project

Sericulture（蚕丝业）is one of the oldest professions developed and practiced by human beings. Weaving as a craft also developed independently in ancient China, India, and Thailand as well as in the Middle East, Europe, Africa, and South America.

The process of working out new techniques continued for millennia（千年）and continues even today. With the change from subsistence farming（自给农业）to the present commercial system, incorporations（公司）of mechanical technology have brought in drastic changes in both sericulture and weaving industry.

Consequently, people realize the present day techniques adopted for commercial system may be unsustainable in the long run.

Work in groups to complete the following work:

1. Why are the present day techniques adopted for commercial system unsustainable? How can we solve this problem in the long run?

2. Please go to some silk museums with your group members for a look at the development of sericulture. How did the weaving techniques evolve from ancient times to today's commercial system?

Unit 3 Keys

Unit 4 Designs and Costumes

Unit Guide: In ancient China, a dragon with five claws was an exclusive pattern only for the emperor and different animals were embroidered on the robes to mark two main groups of officers: civil and military. Embroidery is a handicraft that uses only a needle and a thread but brings together the fusion of traditional culture and modern fashion. In **Passage 1** of this unit, we will explore the figurative language and the symbolic meaning of the various animals, mythological creatures and plants in Chinese silk embroidery; and in **Passage 2**, we will look at how Chinese designers are drawing inspirations from this ancient tradition and bringing new life to this silent heritage.

Lead-in

I Test your knowledge about embroidery. Choose the best answer to each of the following sentences.

<table>
<tr><td>1. Having its beginnings in the _______ age, silk embroidery skills have been handed down through generations.
(A) Bronze
(B) Iron
(C) Paleolithic
(D) Neolithic</td><td></td></tr>
<tr><td>2. _________ embroidery is known for its meticulous stitch work and precision, and an offshoot of this style is the double-sided work that is visually flawless from either side of the base material.
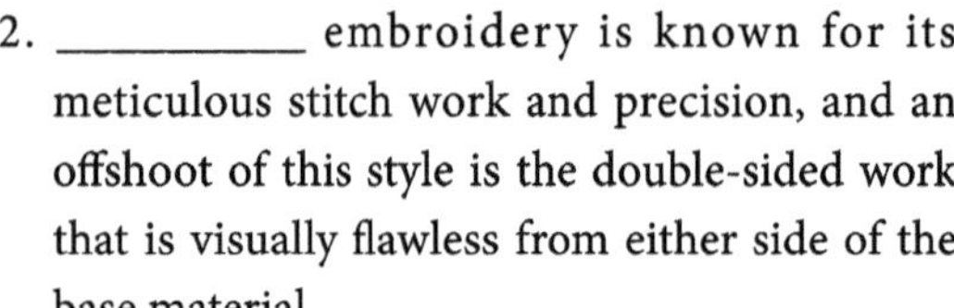
(A) Suzhou
(B) Hunan
(C) Guangzhou
(D) Sichuan</td><td></td></tr>
</table>

3. Within "even embroidery" (ping xiu 平绣), the primary technique is __________ stitching that uses smooth lines of thread in the same length. (A) "gradient stitch" (qiang zhen 抢针) (B) "neat stitch" (qi zhen 齐针) (C) "prodding stitch" (sou he zhen 擞和针) (D) "thickening stitch" (shi zhen 施针)	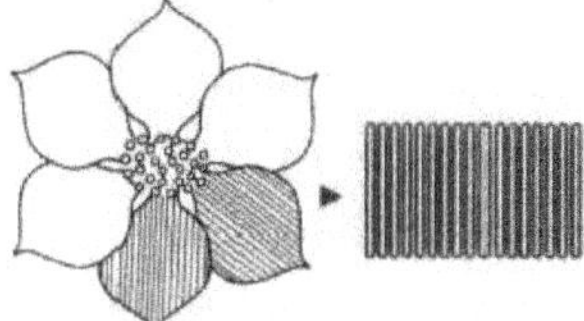
4. Breaking the tradition of "even embroidery," "random embroidery" was developed to use ________ and crossing lines of different colours of thread, layering them to form figures, animals and scenery. (A) oblique (B) neat (C) parallel (D) uniform	
5. Couching consists of using silk threads covered in gold or silver foil laid down side by side, then caught in place by tiny stitches of silk thread. To get unusual effects, these stitches might be in a(an) ________ colour, for instance in using red, the gold would appear to be more copper colour. (A) consistent (B) contrasting (C) similar (D) identical	
6. Different thread thicknesses are used to embroider different subjects. A thinner thread is used for ________, to capture its swift dexterity. (A) stones (B) trunks of trees (C) gold fish bodies (D) tails of goldfishes	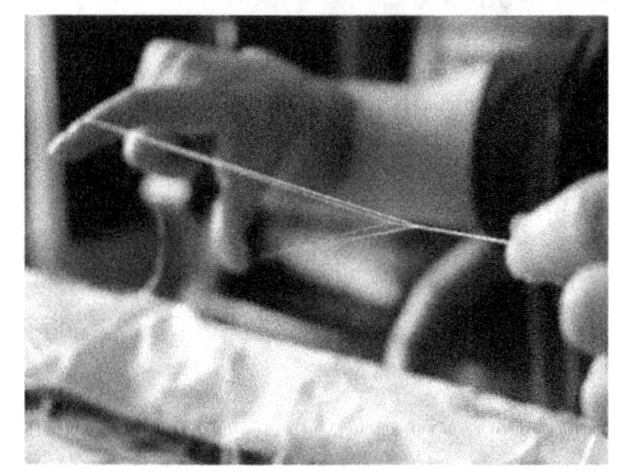

<table>
<tr><td>7. Compared to other traditional embroideries, such as the works from Suzhou and Hunan, Beijing embroidery is known for its rich hued colours, ________ and luxury decorations.
(A) descent patterns
(B) delicate stitches
(C) old-fashioned taboos
(D) special fabrics</td><td></td></tr>
<tr><td>8. In ancient China, royals, especially emperors, used special embroidery to show their status. Thus, taboos developed. For instance, ________ with five claws was an exclusive pattern only for the emperor.
(A) a peacock
(B) a phoenix
(C) a dragon
(D) a tiger</td><td></td></tr>
<tr><td>9. In old China, two groups of men achieved badges, civil and military. Each category had nine ranks. The civil rank badges featured ________ for identification, and the military utilized animals, both real and mythological.
(A) tigers
(B) fishes
(C) birds
(D) dragons</td><td></td></tr>
<tr><td>10. The ________ nature of embroidery is one of the reasons why a younger generation's interest in it has waned.
(A) innovative
(B) aesthetic
(C) entertaining
(D) time-consuming</td><td>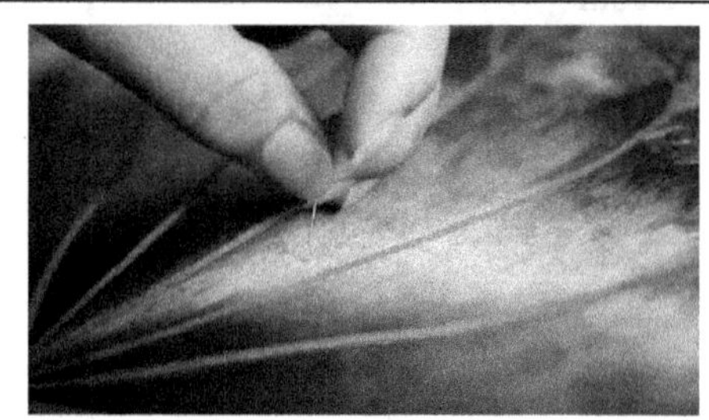</td></tr>
</table>

II A brief talk to the inheritance of traditional embroidery.

Step One: Listen to the recording and fill in the blanks with the exact words you hear.

音 频

It may be a tiny space with each shelf 1.__________ with colourful shoes, but it's not a problem for these customers seeking unique hand-woven designs. The shoes are totally 2.____________. I appreciate this kind of traditional craft. Embroidery is a family 3._________ for Mira Wong. She learned the craft from her grandparents, who set up the business sixty years ago. She's now taken up the 4.________, expanding what's on offer, 5.________ her grandfather's traditional approach.

Many years ago my grandfather only made one style of wearing shoes, but now I've many different styles. And you can see, they're 6.__________ blossoms, dragon and phoenix and also some flowers for the 7._________ of my embroidered shoes. Despite the popularity of shoes, she's worried. The trade is struggling to attract new talent, who are keen to 8.________ it as a long-term career. Many people are just interested in making them as a hobby or making them for their friends and their families, but they won't do it truly as a future career.

It's with that in mind, young 9.___________ are trying to promote this 10. _________ art Mira Wong has embraced social media, earning her an international profile. Others, like Andy GU, are trying to modernize the ancient Chinese designs.

Step Two: A discussion.

The time-consuming nature of embroidery is one of the reasons why a younger generation's interest in it has waned. Studies have shown that with the increasing popularity of electronic forms of entertainment and ability, via digital platforms, to "multi-task," the youth of today may be compromising their attention spans.

Reasons for the wane of being interested in the traditional embroidery	Reasons for the reliance on technology

Ⅲ Chinese master couturier Guo Pei.

Watch the video and answer the following questions.

1. What was Guo Pei's dream when she was young?

____

2. What were the characteristics of the clothes Guo Pei used to wear as a kid?

____

3. What aroused Guo Pei's interest in designing?

____

4. What actually represents Guo Pei's start as a couture designer?

____

5. What did Guo Pei see in the old woman's store in Paris?

____

Passage 1

The Language of Chinese Embroidery

Andrea Stevens Mon

A woman's jacket in the Manchu style dating from the Qing Dynasty Stevens, Andrea. 2015. Auckland War Memorial Museum-Tāmaki Paenga Hira. ID T6.

A gold **embroidered Manchu**-style jacket was one of the remarkable "hidden objects" we displayed during Chinese New Year celebrations this year. To learn more about the garments and the art of Chinese **embroidery**, we invited two experts from the Confucius Institute[1] in Auckland to discuss the **figurative** and **poetic** symbols sewn in silk and gold.

There are many wonderful traditional garments from Asia on display in the Museum, but for Lantern Festival we decided to go down to the basement to find some hidden treasures. We selected a Manchu-style woman's jacket, an exquisite royal garment dating from the Qing Dynasty; a gold embroidered rank **badge**; and two sets of intricately embroidered sleeves, paired and laid flat by a previous collector for display.

Nora Yao and Peter Sun from the Confucius Institute in Auckland interpreted the items at a public talk during Lantern Festival. They explained the figurative language in the embroidery, and the symbolic meaning of the various animals, **mythological** creatures and plants in Chinese history: the plum **blossom** for good fortune, the **crane** for longevity and the **phoenix** for virtue, duty and **resilience**.

This symbolic tradition, used in many Chinese arts, is rich and complex and incorporates a Chinese worldview based on the principles of yin and yang and the five elements[2]. There are many references to **Taoism** and **Confucianism**, and, not least, the poetry and word play within the Chinese language itself.

Chinese hand embroidery dates back to the Neolithic age[3] with production reaching its peak in the 14th century. It is made with fine silk thread—it can be as thin as one 16th the thickness of cotton—displaying wonderful colour and sewn to produce dazzling artistic effects. **UNESCO** cites Chinese embroidery as **Intangible Cultural Heritage**, part of the "traditions or living expressions inherited from our **ancestors** and passed on to our descendants."

These examples from the Qing Dynasty allow us, 11,000 kilometres away from Mainland China, to marvel at their beauty and to see how a different culture interprets the world and their place in it.

In Western legends, the dragon is usually something to be feared. But in Chinese culture, the opposite is true: dragons symbolize protection, good fortune, wisdom, **nobility**, **divinity** and power. Chinese Emperors had them adorn garments and buildings to portray greatness, power and strength. And they continue to be celebrated in Chinese culture today.

This woman's jacket (in the Manchu style) is from the (probably late) Qing Dynasty and has four dragons in total, embroidered in gold and silk thread. "A five-clawed dragon is a royal dragon," explains Peter, "and only the royal dragon can use five claws. There are two phoenix birds depicted also, flying above the **pagoda** on the front of the garment. The phoenix is related to females and, when combined with the dragon, it tells us the garment would be worn by females from the royal family. For an **empress** garment, the dragons would be replaced by phoenixes."

Symbols of Status

Another highlight from the Chinese embroidery collection was this exquisite rank badge[4] (also known as a **Mandarin** square) depicting an embroidered silver **pheasant** against a background of golden clouds, waves and rocks in gold **metallic** thread, with the bird **arching** up towards a red sun disk. The border is a series of couched gold thread **medallions** (held

Symbols for good luck and protection

Front detail of the woman's jacket. Against red silk cloth, the dragon's body is sewn in couched gold thread with white, blue, green and red silk floss used to highlight its detail. Note the five claws of the royal dragon.

Stevens, Andrea. 2015. Auckland War Memorial Museum-Tāmaki Paenga Hira. ID T6.

Front detail. A phoenix flies above the pagoda. The mountains and sea below have a large sea creature breaking through the waves and (just above) a bat (for longevity, happiness, good luck) flies downward next to a plum blossom (for good fortune).

Stevens, Andrea. 2015. Auckland War Memorial Museum-Tāmaki Paenga Hira. ID T6.

A rank badge for a fifth rank civil official, featuring a silver pheasant perched on a rock. Qing Dynasty. Stevens, Andrea. 2015. Auckland War Memorial Museum-Tāmaki Paenga Hira. ID T669.

onto the fabric with a second thread) and backed with blue silk.

"You have probably seen movies set in the Ming or Qing dynasties showing officials wearing a robe with this type of square **insignia** at the front and back," says Peter. "This embroidered badge from the Qing Dynasty has a silver pheasant which **denotes** a fifth rank official, roughly in the middle of the nine-tier ranking system[5], quite a high position."

In her book *Chinese Dress: From the Qing Dynasty to the Present*, Valery Garrett describes how the badges were applied to a centre-fastening **surcoat** for formal court occasions. She notes that the use of rank badges, a Ming custom, was taken up again by the conquering Manchu from about 1652: "With slight modifications, they continued the system of **demarcating** the nine ranks of civil officials[6] by birds embroidered on the squares and the military Mandarins[6] by animals. The ability of birds to fly high to heaven indicated the superiority of the civil mandarins over their military counterparts whose animals were earth-bound."

The silver pheasant, one of the Twelve Symbols of Sovereignty[7] representing Chinese imperial authority, symbolizes the literary ability required for such a high rank. The pheasant is also combined with symbols for longevity to bestow good fortune on the wearer: the clouds are illustrated in a golden never-ending Chinese knot (meaning eternal) and a circular longevity symbol is applied as a border pattern. There is extensive use of couched gold metallic thread making for a very impressive official court badge.

Symbols for a Happy Marriage

Double textile panel

This embroidery panel is made from a pair of sleeves, which appear to have been removed from a garment and laid flat by a previous collector. The soft blues, creams and browns portray a dreamy, imagined landscape with pairings of birds, flowers and insects.

Figuratively, they inspire a happy marriage, with most of the objects in pairs: the elegant **mandarin ducks**, beautiful birds that symbolize **fidelity**; nearby lotuses for purity; two butterflies once again recalling **"The Butterfly Lovers"** ; and two **magpies** chirping amongst the plum blossoms, **heralding** good news.

"Take the lotus flowers on the embroidery piece as an example," says Peter. "The two lotus flowers grow out of the same lotus root. Though separate from each other, they are still linked with each other through fibres of the lotus root. The correlation is usually used to describe the **reconciliation**, **constancy**, purity and harmony of love between husband and wife."

Another whole layer of meaning comes alive for Mandarin speakers of course. "Plum blossom has a similar pronunciation to 'eye brow, ' " explains Peter, "so if a magpie (representing good news) rests on the branch of the plum blossom, it has the meaning 'happiness up to one's eyebrows. '[8]"

The garment serves as symbol, inspiration and artwork. It is a beautiful example of an embroidery with exquisite colour schemes and displaying the maker's skill in depicting animals.

Chinese embroidery today

Nora Yao concluded this fascinating presentation by noting that, like in many cultures, young women were no longer taught to embroider. "Today only the professionals are able to produce clothes like this, and because of the cost of labour they are very, very expensive and usually only worn by the rich. They are hard to wash so are only used on special occasions. It is a dying art for everyday use."

Words

embroidered [ɪm'brɔɪdəd] *adj.* 绣花的；刺绣的

Manchu ['mæntʃuː] *adj.* 满族的

embroidery [ɪm'brɔɪd(ə)ri] *n.* 刺绣；刺绣品；粉饰

figurative ['fɪg(ə)rətɪv] *adj.* 比喻的

poetic [pəʊ'etɪk] *adj.* 诗的，诗歌的；诗意的；诗人的

badge [bædʒ] *n.* 徽章；证章

mythological [mɪθə'lɒdʒɪk(ə)l] *adj.* 神话的；神话学的；虚构的

blossom ['blɒs(ə)m] *n.* 花；花开的状态

crane [kreɪn] *n.* 鹤

phoenix ['fiːnɪks] *n.* 凤凰

resilience [rɪ'zɪlɪəns] *n.* 弹力；顺应力

ancestor ['ænsestə] *n.* 始祖，祖先；被继承人
nobility [nə(ʊ)'bɪləti] *n.* 贵族；高贵；高尚
divinity [dɪ'vɪnəti] *n.* 神；神性；神学
pagoda [pə'gəʊdə] *n.* （东方寺院的）宝塔
empress ['emprɪs] *n.* 皇后；女皇
mandarin ['mænd(ə)rɪn] *n.* 清朝官吏
pheasant ['fez(ə)nt] *n.* 雉科鸟
metallic [mɪ'tælɪk] *adj.* 金属的；含金属的
arch [ɑːtʃ] *vi.* 拱起；成为弓形
medallion [mɪ'dæljən] *n.* 大奖章；圆形浮雕
insignia [ɪn'sɪgnɪə] *n.* 标志；徽章；荣誉
denote [dɪ'nəʊt] *vt.* 表示；指示
surcoat ['sɜːkəʊt] *n.* 外衣
demarcate ['diːmɑːkeɪt] *vt.* 划分界线；区别
fidelity [fɪ'delɪti] *n.* 忠诚；尽责
magpie ['mægpaɪ] *n.* 鹊，喜鹊
herald ['her(ə)ld] *vt.* 通报；预示……的来临
reconciliation [ˌrek(ə)nsɪlɪ'eɪʃ(ə)n] *n.* 和谐；甘愿
constancy ['kɒnst(ə)nsi] *n.* 坚定不移

Proper Names

Taoism 道家的学说；道教
Confucianism 孔子学说；儒家思想
UNESCO 联合国教育、科学及文化组织（简称：联合国教科文组织，英文：United Nations Educational, Scientific and Cultural Organization）
Intangible Cultural Heritage 非物质文化遗产
mandarin ducks 鸳鸯
The Butterfly Lovers 中国古代民间四大爱情故事之一，文中指“梁山伯与祝英台”的故事

Notes

1. Confucius Institute is a non-profit public educational organization affiliated with the Ministry of Education of the People's Republic of China, whose stated aim is to promote

Chinese language and culture, support local Chinese teaching internationally, and facilitate cultural exchanges.
孔子学院，是中国国家汉语国际推广领导小组办公室在世界各地设立的推广汉语和传播中国文化的机构。文中是指位于新西兰奥克兰的孔子学院。

2. Yin and yang in Chinese philosophy, describes how seemingly opposite or contrary forces may actually be complementary, interconnected, and interdependent in the natural world, and how they may give rise to each other as they interrelate to one another. Yin is the receptive and yang the active principle, seen in all forms of change and difference. Wu xing, also known as five elements (Venus—金, Jupiter—木, Mercury—水, Mars—火, Saturn—土) is the short form of "wǔ zhǒng liú xíng zhī qì" (五种流行之气) or "the five types of chi dominating at different times." It is a fivefold conceptual scheme that many traditional Chinese fields used to explain a wide array of phenomena, from cosmic cycles (宇宙周期) to the interaction between internal organs, and from the succession of political regimes (政治体制) to the properties of medicinal drugs (医用药品).
阴阳五行，可分为"阴阳"与"五行"，然而两者互为辅成，五行必合阴阳，阴阳说必兼五行。阴阳五行是中国古典哲学的核心，为古代朴素的唯物哲学。阴阳，指世界上一切事物中都具有的两种既互相对立又互相联系的力量；五行即由"金、木、水、火、土"五种基本物质的运行和变化所构成，它强调整体概念。阴阳与五行两大学说的合流形成了中国传统思维的框架。这种象征性的传统运用于许多中国艺术中，其内涵丰富而复杂。

3. Neolithic Age (also known as the New Stone Age), the final division of the Stone Age, began about 12,000 years ago when the first development of farming appeared in the Epipalaeolithic Near East (旧石器时代末期的近东地区), and later in other parts of the world. The division lasted until the transitional period of the Chalcolithic (铜石并用时代) from about 6,500 years ago (4500 BC), marked by the development of metallurgy (冶金术), leading up to the Bronze Age and Iron Age.
新石器时代是考古学家设定的一个时间区段，大约从12000多年前开始，结束时间距今5000多年。中国手工刺绣可以追溯到新石器时代，在14世纪其生产达到顶峰。

4. A rank badge , also known as a Mandarin square, was a large embroidered badge sewn onto the surcoat of an official in Imperial China. It was embroidered with detailed, colourful animal or bird insignia indicating the rank of the official wearing it.

补子，系补缀于品官补服前胸后背之上的一块织物。为明清官服饰制度的一个重要特征。饰以禽兽纹样来区分官员等级的方法，最早源于唐武则天时期。

5. The nine-tier ranking system, also known as the nine grade controller system, was used to categorize and classify government officials in Imperial China. The nine ranks were separated into upper, middle, and lower classes, each composed of three ranks, making nine in total. Each rank was also further classified into standard and secondary ranks, so that the entire system contained 18 ranks.
九品，指旧时官秩分九等，称为“九品”，所以整个官职系统共有十八个等级。每品又分正从；泛指九个等级。

6. Civil officials and military Mandarins: The Qing Dynasty divided the bureaucracy into civil and military positions, both having nine grades or ranks, each subdivided into primary and secondary categories. Civil appointments ranged from attendant (伺者) to a Grand Secretary (大学士) in the Forbidden City (highest) to being a county magistrate (县令), prefectural (地方政府的) tax collector, and so on. Military appointments ranged from being a field marshal or chamberlain of the imperial bodyguard (侍卫大臣) to a third class sergeant (三等军士), corporal (下士) or a first or second class private (上等士兵).
清朝为加强中央集权，削弱、分化大臣权利，以防权臣篡位，建立了有别于其他各朝的官制，有九品十八个级别，分中央和地方官职两大类。中央官职分中枢部、军机处和帝室部三类，如内阁、军机处、六部等。地方官职分文官、武官等几类，如总督、巡抚、将军、提督等。

7. Twelve Symbols of Sovereignty, also known as Twelve Ornaments, are a group of ancient Chinese symbols and designs that are considered highly auspicious. They were employed in the decoration of textile fabrics in ancient China, which signified authority and power, and were embroidered on vestments of state.
十二章纹，是中国帝制时代的服饰等级标志，帝王及高级官员礼服上绘绣的十二种纹饰，分别为日、月、星辰、群山、龙、华虫（有时候分花和鸟两个章）、宗彝（南宋以前就是一只老虎一只猴子）、藻、火、粉米（晋朝以前是粉和米两个章）、黼、黻。

8. A magpie rests on the branch of the plum blossom, which has the meaning "happiness up to one's eyebrows. " A magpie represents good news, so "happiness up to one's eyebrows" means "喜上眉梢" in Chinese.
一只喜鹊停在梅花枝上，有“眉开眼笑”的意思。喜鹊代表好的消息，所以英文“眉飞色舞”在汉语里就是“喜上眉梢”的意思。

Reading Comprehersion

Choose the best answer to each of the following questions or sentences.

1. Which of the following items is NOT selected on display for Lantern Festival according to the passage?
(A) A Manchu-style woman's jacket.
(B) A gold embroidered rank badge.
(C) A golden court woman's headdress.
(D) Two sets of intricately embroidered sleeves.

2. As the figurative language in the embroidery, the plum blossom denotes ______________.
(A) good fortune (B) longevity
(C) virtue (D) resilience

3. When does Chinese hand embroidery date back to with production reaching its peak in the 14th century?
(A) The Bronze age (B) The Iron age
(C) The Paleolithic age (D) The Neolithic age

4. In Chinese culture, dragons symbolize much EXCEPT ________________.
(A) protection (B) nobility
(C) ferocity (D) power

5. According to the passage, the phoenix embroidered on the front is related to females and, when combined with the dragon, it tells us that ___________ .
(A) the garment symbolizes the social status of middle class
(B) the garment would be worn by females from the royal family
(C) the garment might show a popular embroidery among young women
(D) the garment stands for the aesthetic value in the era

6. What does the ability of birds embroidered on the square to fly high to heaven indicate?
(A) It indicates the prosperity of the civil mandarins in their career development.

(B) It indicates the flourish of the country's political and economic status.
(C) It indicates the success of the frontier guard against alien races.
(D) It indicates the superiority of the civil Mandarins over their military counterparts

7. What does the silver pheasant, one of the Twelve Symbols of Sovereignty symbolize?
   (A) The military ability required for a high rank.
   (B) The literary ability required for a high rank.
   (C) The freedom bestowed by emperor.
   (D) The right bestowed by emperor.

8. The two lotus flowers on the embroidery piece grow out of the same lotus root. Though separate from each other, they are still linked with each other through fibres of the lotus root. What is the correlation used to imply?
   (A) The pure and harmonious love between husband and wife.
   (B) The reconciliation between parents and children.
   (C) The trust between emperor and minister.
   (D) The constancy between master and disciple.

Summary

The following is a summary of Passage 1. Fill in the blanks with appropriate words.

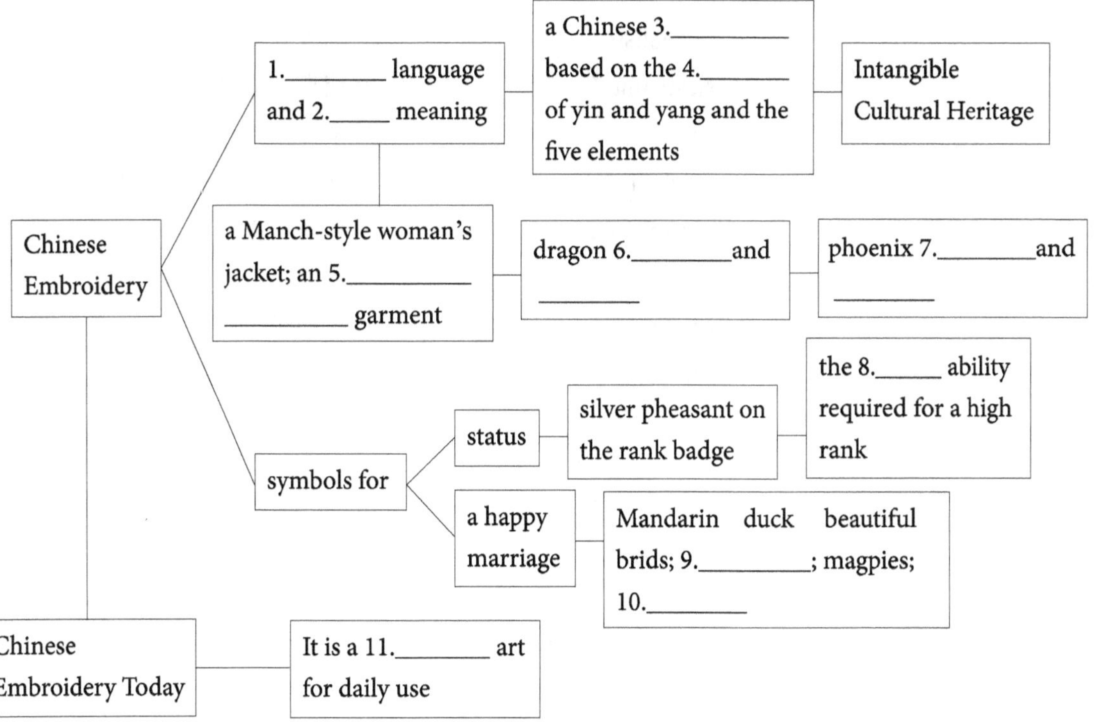

Word Building

I The root *art* can be used together with other words to form new ones, meaning "skill, joint, trick." Match the following words with their parts of speech and Chinese meanings.

	article	艺术的
	artful	手工艺品
n.	artificial	工匠；技工
	artistry	巧妙的；狡猾的
adj.	artisan	人工的；矫揉造作的
	artifact	发音；清楚地讲话
v.	artistic	物品；文章
	articulate	艺术技巧

II The prefix *sur* can be used together with other words to form new words, meaning "over and above." Match the following words with their parts of speech and Chinese meanings.

	surface	登上；超越
	surpass	多余的
	surrealism	临视；看守
n.	surmount	调查；测量
	surplus	幸存；生存
adj.	surcharge	外衣
	surveillance	超现实主义
v.	survive	超过
	survey	附加费
	surcoat	表面

Words and Expressions in Use

I Fill in the blanks with the words given below. Change the form where necessary. Each word can be used only once.

interpret	mythological	blossom	peak	dazzling	nobility
adorn	highlight	exquisite	literary	herald	fascinating

1. Fuzzy language is usually used in ___________ works by experienced writers to obtain unique expression effect and artistic value.

2. Seed dream is to ___________ and bear fruit, the eagle's dream is ready to fly.

3. The vernal equinox (春分) and autumnal equinox (秋分) ___________ the beginning of spring and fall, respectively.

4. Maybe you did not voyage a lifetime to reach the other side; maybe you are climbing a ___________ you had not been able to board.

5. In this sense, this book is incredibly high minded, revealing his ___________ and intellectual commitments to truth.

6. *Two Springs Reflect the Moon* is a(an) ___________ example of Chinese instrumental folk music stemming from the heart of a small-town folk artist.

7. Previously the only way to access people's dreams is that psychologists ask about them after the event and try to ___________ them.

8. Here and there a bit of green or red light is used to ___________ some spots, but they don't offend the eye.

9. Among the twelve, the dragon is the only ___________ beast, and to be born in the Year of the Dragon is regarded as propitious.

10. The vast variety of his knowledge enabled him to ___________ and light up every subjecton which he touched.

11. The Tang Dynasty is the golden age of Chinese poetry, which boasts the most of Li Bai, a most ___________ star in the sky of thc Tang poetry.

12. One of the most ___________ facets of British society is that there seem to be more eccentrics in England than anywhere else.

II Render the following Chinese expressions into English. The first letter of each word is given.

1. 元宵 L__________ F__________
2. 文官 c__________ o__________
3. 细丝线 f__________ s__________ t__________
4. 艺术效果 a__________ e__________
5. 隐藏的宝藏 h__________ t__________
6. 正式的宫廷场合 f__________ c__________ o__________
7. 只在地上的动物 e__________ a__________
8. 中国的手工刺绣 C__________ h__________ e__________
9. 九等官品体系 n__________ r__________ s__________
10. 主权的十二章纹 t__________ s__________ of s__________

Translation

Translate the following paragraph into English.

旗袍源自清代满族女性服饰，被誉为中国传统服饰文化的典范。它不仅在整体造型的风格方面符合中国文化的特点，而且装饰手法也展现着浓厚的东方特质。另外，穿旗袍可以增加形体的修长感，配上中高跟鞋，更可以抬升人体的重心，将东方女性的端庄、典雅和含蓄的美展露出来。因此，旗袍在中国民族服装中独树一帜，久盛而不衰。

Passage 2

Chinese Fashion Designer Guo Pei Goes Global with Army of Artisans

Vivian Chen

Da Jin (magnificent gold) at the Met museum.

Chinese **haute couturier** Guo Pei, who has gained international **acclaim**, draws sophisticated and **affluent** customers at home.

Despite having been in the trade for more than 30 years, Chinese couturier Guo Pei finds herself at the starting line again.

"Many of my friends are planning their retirement, but for me, it's only the beginning," Guo, 50, says at her **Parisian** showroom in **Place Vendôme**, just a day after her July couture show at **Hotel Salomon de Rothschild**.

Guo started her career as a fast fashion designer in the 1980s and opened her own **made-to-order atelier** in 1997. She went on to become one of the few Chinese designers invited by Fédération de la Haute Couture de la Mode[1] as a guest member to take part in **Paris Haute**

Couture Week alongside heritage couturiers such as **Chanel, Dior** and **Valentino.**

Chinese designer Guo Pei goes for gold at Paris Haute Couture Week

Although Guo had already built a strong fan following in China, counting A-listers the likes of Zhang Ziyi and Li Bingbing as clients, she didn't get in the international **spotlight** until 2015 when **Rihanna** attended the Met Ball[2] in one of her designs. The **spectacular**, yellow **fur-trimmed cape** gown with a lavish train decorated with Chinese embroideries took Guo two years to create.

Chinese couturier Guo Pei earned overnight fame when she designed Rihanna's dress for the Met Gala in 2015.

The wide coverage from fashion magazine spreads to internet **memes** earned Guo overnight fame, which unlocked more opportunities and possibilities.

She was invited to show in Paris two years ago, and her creations were featured in exhibitions at **prestigious** institutions such as **New York's Metropolitan Museum of Art** and **Museum of Decorative Arts in Paris.**

Being invited to show at Paris Haute Couture Week as a guest member, Guo says, was a **steep** learning curve.

Guo Pei is known for her delicate details and lavish designs.

"I respect the values of the couture federation as they respect their members' passion, craftsmanship and how they can contribute to the living art–haute couture–in the future," she says. "Being on the official couture calendar has really taken me and my team to new heights. I feel that we are growing with every collection." Following her breakthrough overseas, Guo received a warm welcome from her clients at home.

"Chinese clients who doubted whether my designs could rival the Western designers' were very impressed and proud of my global recognition," she says.

Affluent Chinese customers' growing appetite for sophisticated luxury has fueled the development of home-grown designers, Guo adds.

"Although China's modern fashion industry doesn't have a long history, customers are learning fast," she says. "Their growing knowledge and appreciation of quality designs have kept me going. Without the clients who have grown with me, designers like me won't have the capacity to grow and the possibilities to **prosper**. They have learned to look beyond the price tag and understand the values of couture."

One of her clients was, in fact, the matchmaker for her recent collaboration with **Chopard** for her haute couture autumn/winter 2017 collection. The client asked for a piece of Chopard high jewellery to match a couture wedding gown by Guo Pei for her big day. The client brought Guo and **Caroline Scheufele**—Chopard's artistic director and co-president—together.

At Guo's July haute couture show in Paris, models **donned** Guo's statement couture gowns matched with Chopard's multicarat diamond high jewellery **accentuated** with colourful precious stones such as **rubies** and **emeralds**.

"The big challenge for us was to create our own collection—be it couture or high jewellery that truly **complements** each other," Guo says.

The starting point for Guo was the bright colours and elegant curves of high jewellery. "The jewellery makes a strong statement," she says. "They are **complicated** yet **feminine**. The jewellery really reminds me of my embroidery."

Inspired by the golden age of haute couture, Guo **opted** for classic ball gown **silhouettes** with feminine **asymmetrical** shapes and cut-out details. Echoing the **shimmer** of high jewellery, Guo incorporated metallic fabrics into the collection.

A look from Guo Pei's autumn/winter 2017 couture collection designed in collaboration with Chopard.

Fashion houses create their own fabrics to highlight craftsmanship

The **labour-intensive** details in the embroideries, **floral** appliqués and **embellishments** that Guo's best known for elevated the classic gowns to a whole new level, which suited the taste of the couture clients and high jewellery collectors on the front row.

The **mesmerising** details could not have been achieved without Guo's army of artisans—500 of them to be exact–300 of which are expert embroiders.

Guo has been recruiting artisans since 20 years ago when she started her own atelier called the Rose Studio.

Based on traditional **artisanal** skills such as Su Xiu[3] (special embroideries made famous by artisans from Suzhou) and Gong Xiu[4] (literally palace embroidery, often seen in garments of imperial families in ancient China), Guo has discovered her unique skills.

"The artisans are my most precious assets," says Guo. "I nurtured them from day one. When we first started, we didn't have many references or role models that we could follow. We started from scratch and I'm very proud of the work they do today."

Couture receives injection of vitality and 'street cred'

Karen Mok wearing a magnificent gown designed by Guo Pei.

Da Jin[5] (literally magnificent gold) for example, the dress which took artisans 50,000 hours to make and one of Guo's signature designs that was exhibited at the Met museum, demonstrated the level of craftsmanship in her work.

Despite being a Beijing-based Chinese designer, Guo deliberately didn't highlight her Chinese roots in her designs, but was inspired by her time in the West, such as a visit to an 18th-century **Swiss cathedral** last season, and the Russian princess or **art deco diva** in previous seasons.

"I think Chinese influences are in my blood," Guo says. "It's my design language. I don't want to be labeled as a Chinese storyteller. I think about a global audience."

Guo says she has built a solid foundation and that she's excited about what the future will bring.

"I have faith in my future because of the customers who have grown with me," she says.

"Now that I have gathered my 30 years of experiences and resources, I'm ready to take it further. I'm ready to fly."

Words

haute [əʊt] *a.* 高级；高级的；昂贵的
couturier [kuːˈtjʊərieɪ] *n.* 女装设计师
acclaim [əˈkleɪm] *n.* 欢呼，喝彩；称赞
affluent [ˈæluənt] *adj.* 富裕的；丰富的；流畅的
Parisian [pəˈrɪziən] *adj.* 巴黎的
made-to-order [ˈmeɪdtəˈɔːdə] *adj.* 定做的；完全合适的
atelier [ˈætəlˌjeɪ] *n.* 工作室；画室
spotlight [ˈspɒtlaɪt] *n.* 聚焦
spectacular [spekˈtækjʊlə] *adj.* 壮观的；公开展示的
fur-trimmed [fɜː(r)ˈtrɪmd] *adj.* 有皮毛装饰的
cape [keɪp] *n.* 披肩
meme [miːm] *n.* 模因
prestigious [preˈstɪdʒəs] *adj.* 有名望的；享有声望的
steep [stiːp] *adj.* 不合理的；夸大的；急剧升降的
prosper [ˈprɒspə] *vi.* 繁荣，昌盛；成功
don [dɒn] *vt.* 穿上
accentuate [əkˈsentʃʊeɪt] *vt.* 强调；重读
ruby [ˈruːbi] *n.* 红宝石
emerald [ˈem(ə)r(ə)ld] *n.* 绿宝石
complement [ˈkɒmplɪm(ə)nt] *vt.* 补足；补充
complicated [ˈkɒmplɪkeɪtɪd] *adj.* 难懂的；复杂的
feminine [ˈfemɪnɪn] *adj.* 女性的
opt [ɒpt] *vi.* 选择

silhouette [ˌsɪlʊˈet] *n.* 轮廓
asymmetrical [eɪsɪˈmetrɪkl] *adj.* 不均匀的；不匀称的
shimmer [ˈʃɪmə] *n.* 微光；闪光
labour-intensive [ˈleɪbərɪnˈtensiv] *adj.* 劳动密集型的
floral [ˈflɔːr(ə)l] *adj.* 花的；植物的
embellishment [ɪmˈbelɪʃmənt] *n.* 装饰；修饰
mesmerise [ˈmezməraɪz] *vt.* 使……入迷；使……目瞪口呆
artisanal [ˈɑːtɪzənəl] *a.* 手工艺性的

Proper Names

Place Vendôme 旺多姆广场，巴黎的著名广场之一，位于巴黎老歌剧院与卢浮宫之间
Hotel Salomon de Rothschild 所罗门・罗斯柴尔德酒店，位于法国巴黎
Paris Haute Couture Week 巴黎高级定制时装周
Chanel 法国奢侈品品牌，创始人是Coco Chanel
Dior 法国著名时尚消费品牌，创始人是Christian Dior
Valentino 全球高级定制和高级成衣奢侈品品牌
Rihanna （人名）瑞哈娜
New York's Metropolitan Museum of Art 纽约大都会艺术博物馆
Museum of Decorative Arts in Paris 巴黎装饰艺术博物馆
Chopard （人名）萧邦
Caroline Scheufele （人名）凯若琳・舍费尔
Swiss cathedral 瑞士大教堂
art deco diva 装饰艺术天后

Notes

1. The Fédération de la Haute Couture et de la Mode brings together fashion brands that foster creation and international development. It seeks to promote French fashion culture, where Haute Couture and creation have a major impact by combining traditional know how and contemporary technology at all times. It contributes to bolstering Paris in its role as worldwide fashion capital.
   高级定制和时尚联合会的业务范围十分广泛，包括时装周承办、新兴品牌发展和时尚教育等多项事业。作为每年国际最富盛誉的巴黎时装周的唯一主办方，该协

会负责制定一年中6个巴黎时装周的日程，其中男装、女装和高级定制各2个。

2. The Met Ball, formally called the Costume Institute Gala and also known as the Met Gala, is an annual fund-raising gala for the benefit of the Metropolitan Museum of Art's Costume Institute in New York City. It marks the grand opening of the Costume Institute's annual fashion exhibit.
纽约大都会艺术博物馆慈善舞会（Metropolitan Museum of Art's Costume Institute in New York City，简称Met Gala / Met Ball）于每年的5月初举行，是时尚界最隆重的晚会，每年的慈善晚会红毯部分都被誉为“时尚界奥斯卡”。

3. Su Xiu (苏绣)—Suzhou embroidery is crafted in areas around Suzhou, Jiangsu Province, having a history dating back 2,000 years. It is famous for its beautiful patterns, elegant colours, variety of stitches, and consummate craftsmanship. Its stitching is meticulously skillful, coloration subtle and refined.
苏绣是中国优秀的民族传统工艺之一，是苏州地区刺绣产品的总称，其发源地在苏州吴县一带，现已遍衍无锡、常州等地。清代是苏绣的全盛时期，真可谓流派繁衍，名手竞秀。苏绣具有图案秀丽、构思巧妙、绣工细致、针法活泼、色彩清雅的独特风格，地方特色浓郁。

4. Gong Xiu is also known as Beijing embroidery (京绣) or Jing Xiu , Gongting Xiu in Chinese, literally the palace embroidery. It was originally made for the imperial household. The history of Jing Xiu dates back to the Tang Dynasty when special workshops were established to produce embroidery items for the imperial household. In the Ming and Qing dynasties, the style of Jing Xiu took shape in materials, handcraftsmanship and embroidery patterns. Jing Xiu is noted for rigid standards of counted stitches, symbolic patterns, and second-to-none skills.
宫绣又称京绣、宫廷绣，其字面意思是宫绣。它最初是为皇室制作的。宫绣的历史可以追溯到唐代，当时建立了专门的作坊，为皇室生产刺绣品。明清时期，宫绣的风格在材料、工艺和刺绣图案上都已形成。宫绣以严格的数针标准、象征性的图案和首屈一指的技巧著称。

5. Da Jin was designed by Guo Pei and shown in the “China: Through the Looking Glass” exhibition in the Metropolitan Museum of Art, New York City, United States in 2015. “China: Through the Looking Glass” was an art exhibition exploring the impact of Chinese aesthetics on Western fashion and how China has fueled the fashionable imagination for centuries.
2015纽约大都会时尚盛典的主题是“中国：镜花水月”。此次展览展出了140多件高级定制以及前卫成衣，并将其置于中国传统艺术经典氛围之中，呈现了中国美学对西方时尚的影响。其中，郭培设计的“大金”受到了万人瞩目。“大金”是由24条绣满精美花纹的金丝绣带拼接而成，100个刺绣工人耗时3个月打造，工时长达5万小时。

Reading Comprehension

I Answer the following questions according to Passage 2.

1. How many years has Guo Pei been in the trade of haute couture?

2. When did Guo opened her own made-to-order atelier?

3. What resulted in her getting in the international spotlight?

4. What was the change of Chinese clients who doubted whether Guo's designs could rival the Western designers'?

5. What did the client ask for to match a couture wedding gown by Guo for her big day?

6. Inspired by the golden age of haute couture, what did Guo opt for?

7. Echoing the shimmer of high jewellery, what did Guo incorporate into the collection?

8. What are the most precious assets to Guo?

II Outline the milestones of Guo Pei's development in fashion design in timeline pattern.

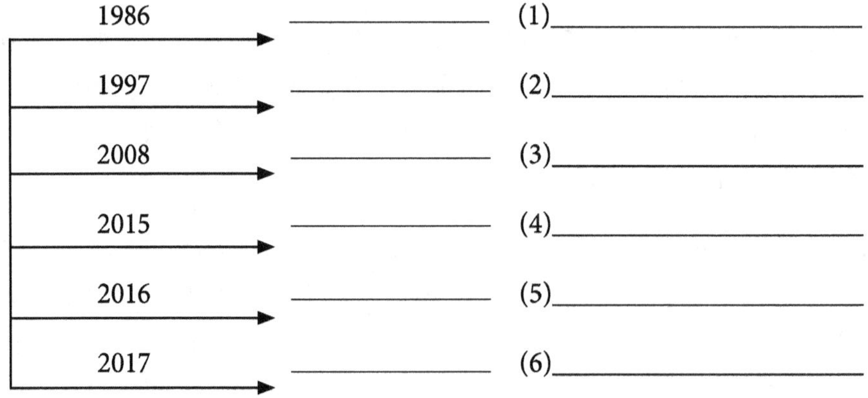

(1)____________________

(2)____________________

(3)____________________

(4)____________________

(5)____________________

(6)____________________

Words and Expressions in Use

I Match the collocations in Column A with the Chinese meanings in Column B.

Column A	Column B
1. gain international acclaim	A. 学得很快
2. count A-listers as clients	B. 以她的作品为重点展出
3. feature her creations	C. 获得国际赞誉
4. a steep learning curve	D. 推动了发展
5. fuel the development	E. 凸显她的中国根源
6. highlight her Chinese roots	F. 把一线明星作为客户

II Compare the following pairs of sentences and explain the different parts of speech and meanings of italicized words.

1. acclaim

(A) But for all its ***acclaim***, the story has never been turned into a feature-length movie.

(B) He's always willing to hear the ***acclaim*** from audiences after he win the game.

(C) Angela Bassett has won critical ***acclaim*** for her excellent performance.

(D) He was ***acclaimed*** as America's greatest filmmaker.

2. count

(A) He was ***counting*** slowly under his breath.

(B) A glass or two of wine will not significantly add to the calorie ***count***.

(C) It's under 7 percent only because statistics don't ***count*** the people who aren't qualified to be in the work force.

(D) No one agrees on what ***counts*** as a desert.

3. steep

(A) San Francisco is built on 40 hills and some are very ***steep***.

(B) Make sure to let the tea ***steep*** a few minutes before taking the tea bag out of the cup.

(C) Consumers are rebelling at ***steep*** price increases.

(D) It's a bit ***steep*** that I should pay for all of you!

4. fuel

(A) They ran out of ***fuel***.

(B) The result will inevitably ***fuel*** speculation about the prime minister's future.

(C) As we like to say in China, many people adding ***fuel*** to bonfire will raise its flame.

(D) The creative spirit ***fuels*** both professional and personal life.

5. complement
   (A) Nutmeg, parsley and cider all ***complement*** the flavour of these beans well.
   (B) There will be a written examination to ***complement*** the practical test.
   (C) The green wallpaper is the perfect ***complement*** to the old pine of the dresser.
   (D) An easy-to-wear shirt dress can ***complement*** all shapes, thanks to its many universally - flattering fit options.

6. highlight
   (A) Last year Collins wrote a moving ballad which ***highlighted*** the plight of the homeless.
   (B) ***Highlight*** the chosen area by clicking and holding down the left mouse button.
   (C) The match is likely to prove one of the ***highlights*** of the tournament.
   (D) It has been a ***highlight*** problem that whether the radiation from the mobile handset may make some side-effect on the human body.

Ⅲ Fill in the blanks with the words or expressions given below. Change the form where necessary. Each word or expression can be used only once.

decorate	overnight	rival	appetite	in the trade	contribute to
look beyond	accentuate	complement	mesmerize	opt for	have the capacity to

1. There is no ________ for learning, particularly amongst senior management, who seem to others to feel that they know it all already.

2. As a university graduate it will be your task to ___________ the present, reach out to a future and define it.

3. In Christianity, verbena is sacred grass, and is often used to _________ in religious consciousness of the altar.

4. Western literary critiques always ________ on the research of literary meaning.

5. The quality of this product is extremely striking and buyers can ________ a number of page layouts and finishes.

6. As an essential and helpful __________ to traditional experiments, the virtual experiment becomes a pop study of experiment teaching.

7. Our company is mainly engaged in pen products production and marketing, and enjoy high reputation __________.

8. Existing infrastructure and facilities ______________ meet the basic needs of cargo passing through the present.

9. Like all great books, it has an enduring power to surprise and ______________ .

10. The area is famous for its wonderfully fragrant wine which has no ___________in the Rhone.

11. Implementation of these measures will give a strong impetus to China's economic development and ______________ world economic growth.

12. If you ignore problems that seem insignificant today, they can blossom into the mother of all crises literally ____________.

Translation Skills

词语英译（4）——汉语成语英译

成语是一种结构固定而凝练、内涵丰富而深刻的语言现象，也是人类的历史和文化在人们所使用的语言中的反映和体现。汉语和英语中都含有丰富的成语。在汉语中，成语是指人们长期以来惯用的、简洁精辟的定型词组或短句，形式以四字居多，也有三字或多字的。而在英语中，成语是词组或句子，其含义往往不等同于所构成单词含义的简单组合。汉英两种语言和文化的巨大差异，也体现在两种语言中成语的表达上。汉语成语形象鲜明、常用典故、音韵和谐的特色在译成英文时，很难全部得以再现，翻译时，应采取恰当的翻译策略，首先确保主要信息的传递。

1. 直译法

在不违背译文语言规范，不引起错误联想的前提下，可以采用直译法，最大限度地保留原文的特色。例如: 临阵磨枪（sharpen one's spear only before going into battle）；狐假虎威（The fox borrows the tiger's terror.）；雨后春笋（like bamboo shoots after a spring shower）；一寸光阴一寸金（An inch of time is an inch of gold.）；雪中送碳（to offer fuel in snowy weather）；一人得道，鸡犬升天（Even the dog swaggers when its master wins favor.）； 花好月圆（blooming flowers and full moon）。这些成语的蕴含意义比较明显，英语读者可以通过译文中词语的意思，了解成语的含义。

2. 意译法

有些成语的蕴含意义很难从字面意义领会到，这时候需要舍弃原文的比喻和形象，采用意译法，传达出原文的蕴含意义。例如：胸有成竹（have a well-thought-out plan before doing something）、开门见山（come straight to the point）、不测风云（Something unexpected may happen anytime.）、黔驴技穷（at one's wit's end）、雕虫小计（insignificant skills）、风马牛不相及（be totally unrelated）。

3. 套译法

东西方民族的思维方式虽然有很大差别，但也有相通之处，有些汉语成语可以在英语中找到喻意相近的表达，这时候就可以采取套译法，直接套用这些现成的英语表达。例如：趁热打铁（strike while the iron is hot）；眼见为实（seeing is believing）；晴天霹雳（a bolt from the blue）；视而不见（close your eyes to something）；绞尽脑汁（rack one's brains）；拒之门外（slam the door in somebody's face）；狂欢作乐（paint the town red）；一心不可二用（A man cannot whistle and drink at the same time.）；智者千虑，必有一失（Homer sometimes nods.）；己所不欲，勿施于人（Do as you would be done by.）；抛砖引玉（throw a sprat to catch a herring）；鱼米之乡（a land of milk and money）；等等。

4. 加注法

加注法是指在译文中添加读者理解成语所需的文化信息和背景知识，特点在于既能保留成语比喻形象的文化蕴含，又能确保译文容易为英语读者所理解接受。例如：守株待兔（watch the stump and wait for a hare—see a Chinese folklore, suggesting waiting for grains without pains）。

Exercising Your Skills

Many Chinese idioms have English Equivalents. Complete the following translations and compare the similarities and differences.

1. 风马牛不相及　　apples and ____________
2. 本末倒置　　Put the _________ before the _________.
3. 王婆卖瓜，自卖自夸　　Every cook praises his own ___________.
4. 人靠衣装　　Fine _________ make fine _____________.
5. 孤立无缘　　out on a _______________
6. 改过自新　　turn over a new ____________
7. 杀鸡取卵　　Kill the __________ that lays the golden _________.
8. 一箭双雕　　to kill two _________ with one __________

Unit Project

The Chinese Clothes—A Combination of Fashion and Classic

One of popular Chinese clothing is the Qipao. Qipao is a familiar style of the traditional dress and includes a tight bodice and a high collar. It is a well-established symbol of Chinese fashion with its elegant and long darling design. The most common fabric for this dress is silk, although other fabrics are possible to match a specific occasion or event. By wearing this type of native dress, it is possible to make a real statement. The style of the traditional dress is made in a variety of styles. The most common methods used to create the unique appearance include the change of the material, the design of the necklace and the length of the skirt. If you are travelling to China on a getaway, there are plenty of nearby tailor shops in major cities that are ready to modify the outfit to match the look and the favoured style.

Streetwear Fashion Is More and More Popular in China

The most visible change in China these two years is about the rising of hip hop culture. The launch of "The Rap of China" has played a decisive role in favour to this trend. China has a booming youth market with new breed of young, cool Chinese Gen-Zs who are putting Chinese street style on the map. "I always hoped that one day our young people would become as cool as the kids in those other cities, shining with their own personalities and styles," Liang explains. "This is what we've been working on: introducing the elements of a streetwear lifestyle with a global vision to young Chinese people and helping them find their own styles." Liang Chao is the founder of Yoho!, a platform that caters to China's urban youth culture.

Work in groups to complete the following work:

1. Qipao stands for the traditional design while streetwear represents the modern preference. Work together to search the internet or library for designs of Qipao or streetwear and present to the class the dress or design you vote to be best. Give the reasons for your choice.

Unit 4 Keys

2. Discuss how to arouse young people's interest in protecting and inheriting traditional design.

Unit 5 Arts and Craftsmanship

Unit Guide: People have been weaving silk into fabric for at least 5,000 years. The delicate material, made from the threads silkworms excrete to create their cocoons, has been used for everything from the robes of Byzantine emperors to the parachutes of World War II paratroopers. But silk is not just a fabric. It is also works of art representing our history and culture. In **Passage 1** of this unit, we will look at the dedication of Li Jing, a young fan maker, to carry on the art and craftsmanship of round fan making; and in **Passage 2**, we will explore the history of Chinese silk painting and appreciate the contributions of modern silk painting masters to this ancient tradition.

Lead-in

I Symbolism in Chinese art.

Chinese people believe that man is an integral part of nature and Chinese artists often use nature as themes of their paintings, embroidery, and so on. As a result, animals, flowers, plants, etc. carry a cultural connotation and became symbols of a definite meaning. For example, cranes often stand for longevity and mandarin duck for faithful love.

Step One: Look at the following pictures and try to figure out what meanings they represent.

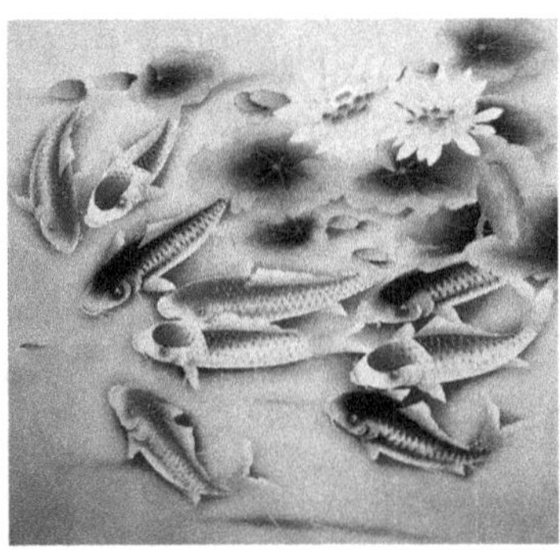

Fish____________________

Lotus____________________

Peony________________

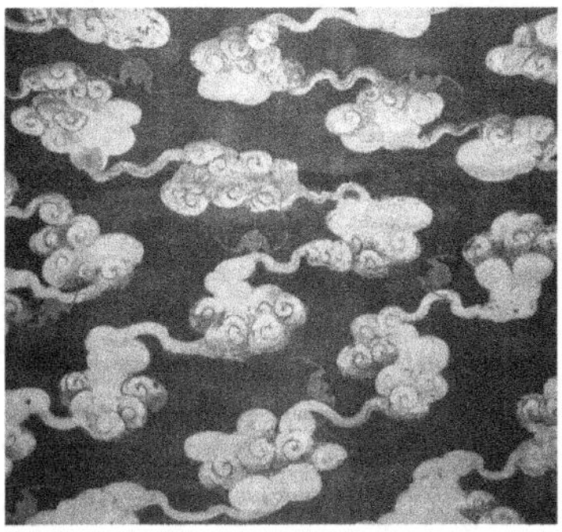

Bat________________

Dragon and Phenix________

Chrysanthemum__________

Plum Blossom__________

Deer________________

Step Two: Work in groups to make a list of other animals, flowers, plants, etc. that often appear in Chinese artworks and explain what symbolic meanings they have.

音 频

II Splendid Chinese handicrafts.

Step One: Listen to the recording and fill in the blanks with the exact words you hear.

1.__________ increasing exchanges between China and other countries and the development of skills, patterns of Western oil paintings and sketches as well as the images of scientific experiments can also now be seen in the embroidery being produced in modern China.

Embroidery epitomizes Chinese people's 2.___________ and artistic sensitivity. Its development is partly based on silk production, which started quite early in ancient China. Chinese painting, which values 3. ___________ and subtlety, also offers many fine patterns to the 4. ___________ , such as mountains, rivers, pavilions, Buddhas, human figures, flowers and birds. This embodies the 5.___________ of various aspects of Chinese culture.

More importantly, as China opens wider to the outside world, the 6. __________ and 7.___________ of Chinese and Western cultures grow rapidly. This has opened up more opportunities and prospects for the 8.___________ of traditional Chinese crafts, and has helped China's 9. ___________ cultural heritage maintain long-lasting 10. ___________ .

Step two: Traditional Chinese crafts have firmly maintained their own charm and popularity even in the face of today's fast, mechanized mass-production techniques. Could you recognize the crafts according to pictures below? If you need to choose one of them to sent to your foreign friends, what is on the top of your list? Why?

These crafts are:

A.________________ B.________________ C.________________
D.________________ E.________________ F.________________

A 	B. 	C. 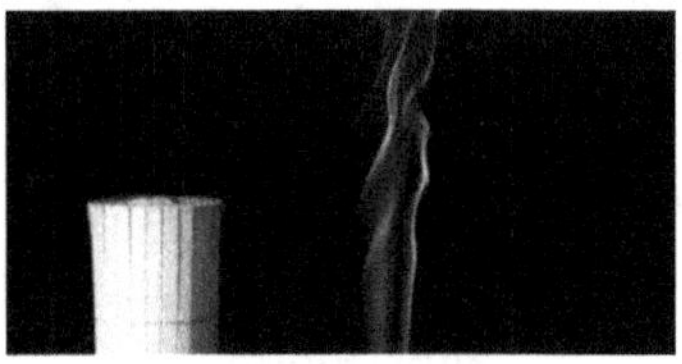
D. 	E. 	F. 

III Exhibition of elaborate Thangka Paintings taking place in Beijing.

Watch a video and fill the blanks with the exact words .

视 频

1. The exhibition of almost 50 masterpieces of 31 Thangka artists from Qinghai Province in northwest China shows cases the ancient art skills that have been__________ __________ from generation to generation.

2. Thangka, Tibetan Buddhist paintings on cotton or silk, usually__________ a Buddhist deity, scene, or mandala.

3. Some of the artworks on display are state-level display __________ __________ __________, like Niangben—a master of arts and crafts and inheritor of Regong arts.

4. The artist can spend years on a single piece, with the ________ and _________ care resulting in many masterpieces.

5. It's commonly believed that at least 10 years of training is required under the __________ __________ of a master to become an accomplished Thangka painter.

6. The _____________ of the skills has remained unchanged, while the number of Tangka painters has expanded quickly.

7. In the past, Thangka artists only passed on their craftsmanship to male ___________, but now they also have female learners.

Passage 1

Reigniting Fan Artisan

Liu Wen. (Photo courtesy: Bazaar)

In 2016, super models and film stars offered some **stunning** fashion shots with the round fans in their hands.

Dressed in a western gown in a British castle, Chinese supermodel Liu Wen[1] still managed to convey the charm and grace of an oriental woman holding a **delicate** fan.

The round fans are from the Leisure Cottage Round Fan Studio in Suzhou, China. The studio is located inside the deep lanes, with a **classical** garden just like the ones in ancient poems and paintings. Li Jing is the owner of the studio. Although graduated from a design school, he wasn't a professional round fan craftsman at the beginning.

Li got to know about the round fans through traditional **operas**. He found them very **appealing** as **props** on stage. Later he began to collect **antiques**, including pieces of round fans. The round fan was actually one of the earliest fans in China. "In the poems and literature works, the round fan is the **epitome** of many traditional Chinese handicrafts, including water and ink painting, embroidery and **inlay**," says Li. "Besides, they are really beautiful." Round fans, i.e., circular fans, were first seen in the Han Dynasty, and became important belongings for women in the palace during the Tang and Song dynasties. Painters and poets liked to leave their works on the fans, turning them into pieces of art. Women tended to hide their faces behind the fans to show their elegance. Sometimes they would use the fans to imply the **solitary** life in the royal palace. All these **infused** the round fans with a sense of feminine grace and sorrow.

Li majored in business administration as an undergraduate but took up art design for his master's degree studies. While in college, he developed a great interest in Peking Opera and had the chance to learn about it from a master named Ye Shenghua.

As part of the process, he gained access to Peking Opera[2] **costumes** and **accessories** and studied the techniques used to make them. The round fan used in traditional operas particularly caught his **fancy**.

"My initial interest was in folding fans, but then I became **obsessed** with Kesi, or Chinese silk tapestry, which is frequently used in round fans," he says. "So my interest turned to that."

At the time, Li was also an **avid** collector of small antiques, and hand-held fans fit nicely into his hobby. Many of the fans he found were damaged, so he taught himself how to do repair work.

Li didn't strictly follow the rule of "restoring the old as the old."[3] Rather, he tried to combine traditional Chinese handicrafts skills with modern methods that strengthened the fans. Kesi silk remained a favorite material.

Kesi is a technique in silk tapestry known for its lightness and clarity of **pattern**. Its name comes from the appearance of cut threads created by the use of color in the designs, which are often pictures copied from classic paintings.

In Kesi works, each color area is woven from a separate bobbin. Therefore, it is a very time-consuming process. There is an idiom saying that "an inch of Kesi is worth an ounce of gold."

The technique first appeared in Suzhou during the Tang Dynasty and survived through the end of Chinese feudal society. Kesi works were prized by men of letters in ancient China. For example, Cao Xueqin, author of one of China's four great **literary** classics, *A Dream of Red Mansions* [4], vividly described a garment made of Kesi silk.

"Kesi is the best for copying calligraphy and paintings because it allows a very smooth change between colors," explains Li. "Besides, Kesi works have the exact same look on both sides, making it ideal for round fans."

Water and ink paintings created from the Song Dynasty are particular favorites in Li's designs. Birds, flowers, fish and insects are the main themes of the paintings.

Li collected all the Kesi silk cloth he loved in a closet. He designed the patterns and trusted the embroidery to Kesi masters in Suzhou. It could take them up to two weeks to do a work, which often made him impatient.

"Sometimes I couldn't sleep well during the waiting time," he says. "It felt like waiting to see a lover, and I kept wishing that time would pass faster."

Apart from Kesi, Li also uses such materials as yarn and ordinary silk to make fans. The key point is how the pattern is designed.

"It is not the case that the more complicated the pattern, the more beautiful the fan," he notes. "The composition of the pattern decides if the fan goes well with the frame and **ornamentation**."

The frame, **rib** and pendant of a fan are equally important to Li. His aim is to create a harmonious piece of work. This is where Li's antique collection comes into play.

Take the **pendants** for example. Li tried different ornaments from his **collection**, including silverware, jade and ox horn. To him, there is nothing more beautiful than a silk fan surface with an antique pendant and a rib made from mottled bamboo.

"I never care about the cost," he says. "I just want to make the most beautiful fan I can. It's all about passion and love."

The artistry of his work has not gone unnoticed. In Li's online store on Taobao.com, China's biggest e-commerce platform, his fans are priced anywhere from 300 yuan (US$45) to 28,800 yuan.

"My customers usually have the same **obsession** as I do," he says. "And they have the money to **indulge** their passions. It is **gratifying** to find people with whom I can share an appreciation of the art of round fans."

Despite such outward appearances of acceptance, Li worries that the art of hand-making round fans will eventually die out if younger generations don't answer the call of history and abandon ancient techniques.

For Li, making fans is like communicating with the ancient people. The enjoyable hobby, however, has to meet the reality: "As craftsmen, we hope that more people could appreciate its beauty and be willing to spend money on it. The art can be passed on, but not merely by **preaching** the love for it. "

Words

reignite [ˌriːɪg'naɪt] *v.* 再点火；再点燃；重新激起
stunning ['stʌniŋ] *adj.* 令人吃惊的
delicate ['delɪkət] *adj.* 微妙的；精美的；纤弱的；易损的
classical ['klæsɪk(ə)l] *adj.* 古典的
opera ['ɒprə] *n.* 歌剧
appealing [ə'piːlɪŋ] *adj.* 吸引人的；动人的；引起兴趣的
prop [prɒp] *n.* 道具
antique [æn'tiːk] *n.* 古董
epitome [ɪ'pɪtəmi] *n.* 典型的人或事物
inlay [ɪn'leɪ] *n.* 镶嵌物；镶嵌细工
solitary ['sɒlətri] *adj.* 独自的；独立的；单个的；唯一的；隐居的
infuse [ɪn'fjuːz] *v.* 灌输
costume ['kɒstjuːm] *n.* 服装；装束；戏装；剧装
accessory [ək'ses(ə)ri] *n.* 配饰
fancy ['fænsi] *n.* 喜爱；幻想；想像力
obsessed [əb'sest] *adj.* 着迷；着迷的
avid ['ævɪd] *adj.* 贪婪的；热心的
pattern ['pæt(ə)n] *n.* 模式；图案；典范；式样
literary ['lɪtərəri] *adj.* 文学的；书面的；精通文学的
ornamentation [ˌɔː(r)nəmen'teɪʃ(ə)n] *n.* 装饰；装饰物
rib [rɪb] *n.* (针织物的)凸条纹，罗纹
pendant ['pendənt] *n.* (match) 配对物；匹配物；附属物；补充物
collection [kə'lekʃn] *n.* 收藏品
obsession [əb'seʃn] *n.* 痴迷；困扰
indulge [ɪn'dʌldʒ] *vt.* 满足；纵容
gratifying [grætɪfaɪɪŋ] *adj.* 悦人的；令人满足的
preaching ['pritʃɪŋ] *v.* 讲道 (preach的ing形式)

Notes

1. Liu Wen, born January 27, 1988, is a Chinese model. In 2012, *The New York Times* dubbed her "China's first bona fide supermodel." She is the first model of East Asian descent to

walk the Victoria's Secret Fashion Show, the first spokesmodel of East Asian descent for the Estée Lauder Companies, and the first Asian model to ever make *Forbes* magazine's Annual Highest-paid Models List. In 2017, Liu became the first Chinese model to ever appear on the front cover of *American Vogue*.
刘雯出生于1988年1月27日，中国模特。2012年，《纽约时报》称她为“中国第一位真正的超模”。她是第一位走上“维多利亚的秘密”内衣秀的东亚裔模特，是雅诗兰黛公司的第一位东亚裔代言模特，也是第一位登上《福布斯》杂志年度高薪模特榜的亚洲模特。2017年，刘雯成为第一位登上《美国时尚》封面的中国模特。

2. Peking Opera or Beijing opera (京剧) is a form of Chinese opera which combines music, vocal performance, mime, dance and acrobatics. It arose in the late 18th century and became fully developed and recognized by the mid-19th century. The form was extremely popular in the Qing Dynasty court and has come to be regarded as one of the cultural treasures of China. Major performance troupes are based in Beijing, Tianjin and Shanghai.
京剧，曾称平剧，融唱、念、做、打于一体的戏剧表演形式，在18世纪晚期开始萌芽，并在19世纪中期得以充分发展和广为人知。京剧孕育于民间，融合了中国南北方戏剧元素的京剧，在北京发展成熟，广泛流布于全国。场景布置注重写意，腔调以西皮、二黄为主。京剧在清朝时期颇为流行，用胡琴和锣鼓等伴奏，被视为中国国粹，中国戏曲三鼎甲“榜首”。主要的戏班集中在北京、天津和上海。

3. the rule of “restoring the old as the old” 指古迹等修缮的原则：修旧如旧即按照原有的旧有的样子修缮，修完后面貌与原有设计面貌相同；补新以新指用新的方式对新创作品补充完善，使之更加出色。

4. *A Dream of Red Mansions*, composed by Cao Xueqin, is one of China's four great literary classics. It was written sometime in the middle of the 18th century during the Qing Dynasty. Long considered a masterpiece of Chinese literature, the novel is generally acknowledged to be the pinnacle of Chinese fiction.
《红楼梦》，中国古代章回体长篇小说，又名《石头记》等，被列为中国古典四大名著之首，

一般被认为是清代作家曹雪芹所著。小说以贾、史、王、薛四大家族的兴衰为背景，以富贵公子贾宝玉为视角，描绘了一批举止见识出于须眉之上的闺阁佳人的人生百态，展现了人性美和悲剧美，可以说是一部从各个角度展现女性美的史诗。

Reading Comprehension

Choose the best answer to each of the following questions or senterces.

1. According to this passage, we can conclude that in the ancient times, a woman might convey __________ by holding a delicate round fan.
   (A) her happiness and sadness
   (B) her elegance and grace
   (C) her willingness to marry
   (D) her permission to do something

2. How did Li get to know about the round fans?
   (A) Through his friends.
   (B) Through traditional operas.
   (C) Through the introduction of round fans.
   (D) Through newspapers.

3. Round fans, were first seen in the _______ Dynasty, and became important belongings for women in the palace during the _______ and ______ dynasties.
   (A) Tang, Song, Yuan
   (B) Yuan, Ming, Qing
   (C) Chun, Qiu, Han
   (D) Han, Tang, Song

4. Li tried to combine traditional Chinese handicrafts skills with modern methods that strengthened the fans by ________________________________.
   (A) partly following the rule of "restoring the old as the old."
   (B) deliberately following the rule of "restoring the old as the old."
   (C) strictly following the rule of "restoring the old as the old."
   (D) unwillingly following the rule of "restoring the old as the old."

5. The following are the advantages of the Kesi technique EXCEPT _______________.
   (A) its lightness and clarity of pattern

(B) offering the best way for copying calligraphy and paintings
(C) saving time and energy
(D) its smooth change between colours

6. According to Li, ______________________ decides if the fan goes well with the frame and ornamentation.
(A) complicated pattern
(B) the material of the fan
(C) the sophisticated skill
(D) the composition of the pattern

7. By creating a harmonious piece of work, Li treats the________, ________ and ________ of a fan equally important.
(A) colour, pattern, material
(B) frame, rib, pendant
(C) colour, rib, pendant
(D) frame, rib, material

8. What does the sentence "Younger generations don't answer the call of history." mean?
(A) Younger generations should read more books on history.
(B) Younger generations should call often.
(C) Younger generations may change their tastes and abandon ancient techniques.
(D) Younger generations may forget whom they are.

Summary

Fill in the blanks with appropriate words.

Round fans were first seen in the Han 1.________, and became important 2.________ for women in the palace during the Tang and Song dynasties. Painters and poets liked to 3.________ their works on the fans, 4.________ them into pieces of art. Women tended to 5.______ their faces 6.______ the fans to show their 7.______. Sometimes they would use the fans to 8.______ the solitary life in the royal palace. All these 9.________ the round fans with a sense of feminine grace and sorrow. To some degree, round fans are the 10.________ of many traditional Chinese handicrafts, including water and ink paintings, embroideries and inlays.

Word Building

I The suffix *ant* can be used together with verbs or nouns to form new adjectives or nouns, meaning "... a person or thing that" "that is or does sth." Match the following words with their Chinese meanings.

Column A	Column B
1. tolerant	A. 辅助的
2. assistant	B. 上升的
3. ascendant	C. 能容忍的
4. applicant	D. 申请人
5. accordant	E. 被告
6. occupant	F. 居住者；占有人
7. defendant	G. 和谐的；一致的
8. pendant	H. 部件；成分
9. distant	I. 遥远的；冷漠的
10. component	J. 垂饰；挂件

II The suffix *ship* can be used together with other words to form new words, meaning "quality, condition, state, office, skill" or "the act or skill." Match the following words with their parts of speech and Chinese meanings.

Column A	Column B
1. friendship	A. 成员
2. relationship	B. 友谊
3. membership	C. 关系
4. citizenship	D. 独裁
5. horsemanship	E. 骑马术
6. hardship	F. 苦难
7. dictatorship	G. 合伙关系
8. partnership	H. 推销术
9. airmanship	I. 手艺；技艺
10. salesmanship	J. 飞行技术
11. craftsmanship	K. 居民权

Words and Expressions in Use

I Fill in the blanks with the words given below. Change the form where necessary. Each word can be used only once.

accessory	avid	reignite	stunning	delicate	classical
pattern	obsessed	indulge	gratifying	pendant	ornament

1. He was intensely eager, indeed ________ for wealth.

2. Emperor Qianlong, the sixth emperor of the Qing Dynasty (1644—1911), was _______ with Suzhou cuisine.

3. _____________ are articles such as belts and scarves which you wear or carry but which are not part of your main clothing.

4. The French art of living is present in all 188 rooms and suites and each with a unique decor that combines ________ style and calm.

5. Recalling his multiple meetings with Putin, President Xi says their efforts in guiding a steady and long-term growth of China-Russia ties at a high level have achieved ________ results.

6. Their passion was _________ by a romantic trip to Venice.

7. Inspired by a set of exquisite tea wares unearthed from the 600-year-old Famen Temple in Xi'an, Shaanxi Province, the show recounts the story of a ________ beautiful woman with a farmer, an officer and a monk respectively.

8. It is enjoyable to see something from the larger world reduced to a small, ________ piece of art.

9. Nor should we ______ in self-importance that our culture is the best. As a matter of fact, cultures should be regarded as equal.

10. But for Tiffany's ______ in the shape of Hongbao is less desirable, because the jewelry maker's brand association doesn't resonate with the red envelope's cultural connotations.

11. By the Tang Dynasty, literati were wearing swords as well, as an ________, a sign of standing or of ambition.

12. An ideal ink stone should have an elegant shape and _______, craftsmanship should be top-notch, and materials should be superb.

II Render the following Chinese expressions into English. The first letter of each word is given.

1. 古董收藏 a________ c________
2. 赢得某人的青睐 c________ one's f________
3. 对……着迷 become o_______ with
4. 放纵某人的激情 i________ one's p________
5. 令人惊叹的时尚照片 s________ f_______ shots
6. 耗费时间的过程 t________ p_______
7. 分享对……的欣赏 share an a_________ of
8. 对……产生（培养出）浓厚的兴趣 d______ a great i_____ in
9. 女性的优雅与悲伤感 a sense of f_______ grace and s_________
10. 中国传统工艺品的缩影 the e_____ of the t_____ Chinese h______

Translation

Translate the following paragraph into English.

西湖绸伞始创于1932年，由杭州都锦生丝织厂研创。西湖绸伞具有浓郁的地方特色，用杭州本地独有的淡竹做伞骨，用杭州丝绸做伞面。伞面上的装饰图案有西湖风景，也有国画花鸟、仕女等。西湖绸伞设计奇巧，制作精细，既实用，又有艺术美学价值。2008年，西湖绸伞被列为国家级非物质文化遗产项目。

Passage 2

Chinese Silk Painting: Its History and Spread to the West

Amanda Mailer & Shengfei Zhu

What is Silk Painting?

Originating from China, silk painting (丝绸画) is an art form with over 2,000 years of history that **involves** applying coloured **pigment** to silk cloth. Like its silk embroidery **counterpart**, silk painting **preceded** the invention of paper.

The ancient art was once known as "Bo" (帛) painting, referring to the white silk used as a surface. **Compared to** wood, stone or bamboo of the time, silk was the ideal canvas for painting. It was **luxurious** and yet easy to cut to any desired shape and light to carry.

Chinese artisans of the old first prepared silk cloths by beating it against stone to smooth the surface before applying colour. Using animal hair paintbrushes, ink mixtures of **soot** and glue, or mineral pigments of **vermilion**, **azurite** and **malachite**, ancient Chinese artists created works of art that have survived centuries.

Today silk paintings can be found all around the world, using a combination of dyes and techniques developed in Europe and Asia. The silk surface is often prepared by **stretching** and dying silk with a background colour. Because pigments spread freely when applied to silk, the artist relies less on brushes and more on creating boundaries for the pigment through the use of a resist. Gutta (a rubbery cement) and water-based resists are popular for sketching the outlines of designs on the silk. Once the outlines have dried, dyes are applied to the silk that spread up to the resist borders. **Alternatively**, the silk surface may **be primed to** reduce the dyes' ability to bleed.

In this way silk painting differs to painting on cotton canvas or paper. The artist needs not only to consider the placement of pigment but also control its movement. Similar to watercolour, unrestricted, the ink or dyes will flow freely on the silk, creating soft and **diffuse** artwork.

History and Development of Chinese Silk Painting

Silk painting[1] in China is believed to date back as far as **the Warring States Period** (476 BC—221 BC), reaching its height as an art form in **the Western Han Dynasty** (206 BC—AD 25).

Artisans of the imperial courts first used silk as a medium for calligraphy painting, which at the time was thought to be the highest and purest form of painting. They used black ink made of pine soot and animal-based glues to paint silk **scrolls.**

Over the years the art developed to include human figures and **depict** religious and mythological characters as well as the forms from nature. The oldest silk painting artefacts were unearthed from a tomb built in the Warring States period in Changsha, Central China. The two silk paintings that were discovered featured **mythical** beasts—the dragon and phoenix traditionally believed to help the dead enter heaven.

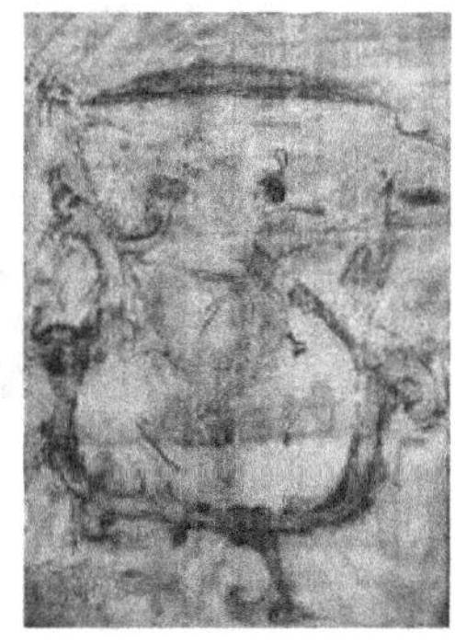

The first painting, titled "Lady, Dragon and Phoenix" depicts a noblewoman on a boat praying to a dragon and phoenix. The second, titled "Man Driving the Dragon" features a bearded nobleman **escorted** by a dragon and an egret. A sacred bird of ancient China, the **egret** is thought to represent the **integrity** and noble qualities of the man.

Both silk paintings show their characters in **profile**, a style typical for the **Chu** people of that period. They are thought to be **burial** items that may have accompanied a funeral procession as banners and were laid with the deceased to protect their souls and help them ascend to heaven.

Until the 2nd century AD silk painting **was exclusive to** China as a result of their efforts to keep sericulture and silk production a secret. As silk became a highly **coveted** trading commodity, the art form gradually spread across Asia, making its way to Europe.

Spread of Silk Painting from China

Legend has it that around 100 AD, a Chinese princess promised to a Khotan[2] prince in

Central Asia smuggled silkworm eggs and mulberry seeds out of the country. In so doing she revealed the secret of silk production and ended China's monopoly on the luxury fabric. Around this time too, silk painting could be found in India where wax was used as a resist in their silk designs.

Around 300 AD, the Japanese came to learn the methods of silk production from the Chinese. Their early silk paintings were **monochromatic**, using black ink and paint. It wasn't until 1300 AD that Japanese artists began to use a range of coloured pigments.

It wasn't until the 12th century, did silk production begin to spread to Western Europe. New manufacturing techniques saw silk production boom, and Italy established itself as a major European centre for silk. By the 18th century, **the industrial revolution** made the cloth even more widely available, and with it spread silk painting as an art form.

In Indonesia, family members of the Russian Tsar, Nicholas Ⅱ[3] learned the **batik method** of silk painting using wax resists. They brought the art to France where the **serti technique** was introduced in the 1900s. This technique of painting using **gutta** resists to control dyes on silk is one of the most widely used today. Silk painting as an art continued to spread and gained popularity in Britain and America by the 1970s.

Today **a multitude of** silk painting styles and techniques abound, and the art can be found all around the world.

Modern Chinese Silk Painting Masters

Currently an art professor at Suzhou University, Wenzheng Dong is considered a pioneer of modern silk painting. He developed a new technique of silk painting, called "Wumo" or "Silent in Suzhou" painting (吴默画)[4]. Applying Western principles of light and contrasting colours to traditional Chinese landscape **compositions**, the Wumo style of silk painting achieves dramatic effects. Wenzheng Dong's artwork has been well received in over 25 countries across Europe, North America and Asia, winning acclaim as a special article of Sino-Franco cultural exchange.

Another highly accomplished artist, Wenjin Bi studied classical Chinese painting and is **accredited** with developing more than 70 silk painting pigments. These colours are praised for their ability to **accommodate** all kinds of silk textures, making them suitable to a wide range of painting styles.

Bi Wenjin however **is renowned for** his mastery of "vermilion-azurite painting" (丹青觉法)[5], using the traditional mineral pigments to **enliven** the dark ink of classical Chinese silk painting. He is most famous for his paintings of **chrysanthemum**, and their expression of traditional Chinese aesthetics.

Silk Painting Today

Silk painting has come a long way from its beginnings in the ancient imperial courts. Smuggled to Central Asia, seized by the Crusades and combined with batik inspired methods in Western Europe and brought to America, silk painting is now practiced all around the world. Modern–day techniques, such as Seri, with its simple outline and dye method, have made silk painting as easy as filling in a colour book and accessible to all.

Meanwhile in China the art continues to develop, adopting painting techniques from the West and renewing classical bird and flower styles of the old. As silk painting **evolves** into a modern art form, we can be sure it will **flourish** for generations to come.

Words

involve [ɪnˈvɒlv] *vt.* 包含；牵涉

pigment [ˈpɪgmənt] *n.* 颜料；色料

counterpart [ˈkaʊntəpɑːt] *n.* 极其相似的人或物

precede [prɪˈsiːd] *vt.* 在……之前发生或出现；先于

luxurious [lʌgˈʒʊəriəs] *adj.* 豪华的；奢侈的

soot [sʊt] *n.* 烟灰；油烟

vermilion [və(r)ˈmɪliən] *n.* 朱红色；鲜红色

azurite [ˈæʒʊraɪt] *n.* 石青；蓝铜矿

malachite [ˈmæləkɪt] *n.* 孔雀石

stretch [stretʃ] *v.* 伸展；延伸；持续；包括

alternatively [ɔːlˈtɜːnətɪvli] *adv.* 或者；二者择一地；要不然

diffuse [dɪˈfjuːz] *adj.* 冗长的；四散的

scroll [skrəʊl] *n.* 画卷

depict [dɪˈpɪkt] *vt.* 描述；描绘，描画

mythical [ˈmɪθɪkl] *adj.* 神话的；虚构的

escort [ˈeskɔːt] *v.* 护送；护卫

egret [ˈiːgrət] *n.* 白鹭；鹭鸶

integrity [ɪnˈtegrəti] *n.* 正直，诚实

profile [ˈprəʊfaɪl] *n.* 侧面，半面；外形，轮廓

burial [ˈberiəl] *n.* 葬礼；葬，掩埋

coveted [ˈkʌvɪtid] *adj.* 令人垂涎的；垂涎的，梦寐以求的

monochromatic [ˌmɒnəkrəˈmætɪk] *adj.* 单色的

batik [ˈbætɪk] *n.* 蜡防印花法

gutta [ˈgʌtə] *n.* 滴；水滴

composition [ˌkɒmpəˈzɪʃn] *n.* 创作；构图；布置

accredit [əˈkredɪt] *v.* 归因于；委托，授权；相信；认可

accommodate [əˈkɒmədeɪt] *v.* 容纳；使适应；向……提供住处；帮忙

enliven [ɪnˈlaɪv(ə)n] *vt.* 使活泼；使生动；使有生气

chrysanthemum [krɪˈsænθəməm] *n.* 菊花

evolve [ɪˈvɒlv] *v.* 使发展；使进化；设计；制订出；发出；散发

flourish [ˈflʌrɪʃ] *v.* 挥舞；茂盛；繁荣；活跃；蓬勃

Useful Expressions

compared to　与……相比，跟……相比
be primed to　准备就绪
be exclusive to　专为……所独有
a multitude of　大量
be renowned for　闻名于世

Proper Names

the Warring States Period (476 BC—221 BC)　战国时期
the Western Han Dynasty (206 BC—AD 25)　西汉时期
Chu　中国战国时期楚国
the industrial revolution　工业革命
batik method　蜡染法
serti technique　蜡染色法

Notes

1. The silk paintings of the State of Chu in the Warring States Period of China are the earliest paintings made of white silk, which were unearthed in the Chu area in the middle and late Warring States Period. The two paintings preserved intact are unearthed in Changsha, Hunan Province.
   《龙凤仕女图》（*The Portrait of A Lady with Dragon and Phoenix*）和《人物御龙图》（*The Portrait of Man Driving the Dragon*）是中国战国时期楚国帛画，是迄今发现时代最早的以白色丝帛为材料的绘画，出土于战国中晚期楚地，保存完整的是湖南长沙出土的两件。

2. The Kingdom of Khotan was an ancient Iranian Saka Buddhist kingdom located on the branch of the Silk Road that ran along the southern edge of the Taklamakan Desert in the Tarim Basin (modern Xinjiang, China). The ancient capital was originally sited to the west of modern-day Hotan (和田) at Yotkan. From the Han Dynasty until at least the Tang Dynasty it was known in Chinese as Yutian (于阗). This large Buddhist kingdom existed for over a thousand years until it was conquered by the Muslim Kara-Khanid Khanate in 1006, during the Islamicisation and Turkicisation of Xinjiang.
   于阗国是古代西域佛教王国，中国唐代安西都护府安西四镇之一。君主国姓为尉

迟。国祚长达1238年。古代居民属于操印欧语系的吐火罗人。1006年被喀喇汗国吞并，逐渐伊斯兰化。11世纪，人种和语言逐渐回鹘化。于阗地处塔里木盆地南沿，以农业、种植业为主，是西域诸国中最早获得中原养蚕技术的国家，故手工纺织发达。特产以玉石最有名。于阗自2世纪末佛教传入后，逐渐成为大乘佛教的中心，魏晋至隋唐，于阗国一直是中原佛教的源泉之一。

3. Nicholas Ⅱ (18 May 1868—17 July 1918), known as Saint Nicholas the Passion-Bearer in the Russian Orthodox Church, was the last Emperor of Russia, ruling from 1 November 1894 until his forced abdication on 15 March 1917. His reign saw the fall of the Russian Empire from one of the foremost great powers of the world to economic and military collapse. He was given the nickname "Nicholas the Bloody" or "Vile Nicholas" by his political adversaries due to the Khodynka Tragedy, anti-Semitic pogroms, Bloody Sunday, the violent suppression of the 1905 Russian Revolution, the execution of political opponents, and his perceived responsibility for the Russo-Japanese War (1904–1905).
尼古拉二世·亚历山德罗维奇（1868年5月18日—1918年7月17日），在俄罗斯东正教中被称为圣尼古拉斯，是俄罗斯帝国末代皇帝（1894年11月1日—1917年3月15日在位），俄罗斯罗曼诺夫王朝最后一位沙皇。在他的统治下，俄罗斯帝国从最重要的大国之一走向了经济和军事的崩溃。由于霍廷卡悲剧、反犹太人大屠杀、血腥星期日、1905年俄国革命的暴力镇压、对政治对手的处决，以及他对俄日战争（1904—1905）的责任，他的政治对手给他起了一个绰号"血腥的尼古拉斯"或"卑鄙的尼古拉斯"。

4. Wu Mo silk painting was created by the famous Chinese painter Mr. Dong Wenzheng. Inspired by ancient Chinese silk paintings, Wu Mo paintings are drawn on the silk crepe de Chine fabric known as the Queen of Fibers. They combine freehand brushwork and fine brushwork of Chinese paintings to absorb the creative skills of Western Impressionists and produce brilliant artistic effects of light and color.
吴默画（"Wumo" or "Silent in Suzhou" painting）是由中国著名画家董文政先生创立的绘画技法，又名吴默帛画。吴默画受中国古代帛画启发，在被誉为纤维皇后的真丝双绉面料上，用中国画的写意泼彩和工笔精绘，吸收西方印象派画家的创作技巧，产生光与色的斑斓艺术效果。

5. Vermilion–azurite painting (Danqingjue), an ancient painting method, was revived by Mr. Bi Wenjin. Mr. Bi developed and sorted out the ancient Chinese painting technique after decades' efforts. Its core is wet standing painting. Freehand brushwork in traditional

Chinese painting is characterized by vivid expression and bold outline. "Danqingjue Method" perfectly embodies the texture characteristics of art carriers through a process of sinking, soaking, moistening and dyeing, i.e. the "Four Principles in traditional Chinese painting."

毕文瑾先生历经多年的研究，发展并整理了沉积了2000多年的中国古老的绘画技法——丹青觉法。其核心是湿立而绘，工笔以矾彩勾青，写意以水没丹青（此"没"非墨）。丹青觉法通过洇、浸、润、染"四鉴之理"绘制的作品，完美地体现了载体艺术的肌理特性。

Reading Comprehension

I Read Passage 2 again and decide whether the following statements are T (true) or F (false).

1. (　　　　) Silk painting came before the invention of paper.

2. (　　　　) Though expensive, silk was still the ideal canvas for painting compared with wood, stone or bamboo of the time.

3. (　　　　) Because pigments spread freely when applied to silk, the artist relies more on brushes and less on creating boundaries for the pigment through the use of a resist.

4. (　　　　) As to silk painting, artists only need to control the pigment's movement because silk painting is similar to paint on cotton canvas or paper.

5. (　　　　) Silk painting, started in the Warring States Period (476 BC—221 BC) , reached its height as an art form in the Western Han Dynasty (206 BC—AD 25).

6. (　　　　) The oldest silk painting artifacts were excavated from a tomb built in the Eastern Han Dynasty in Changsha, Central China.

7. (　　　　) Until the 2nd century silk painting was exclusive to China as a result of their efforts to keep sericulture and silk production a secret.

8. (　　　　) It is believed that a Khotan prince in Central Asia smuggled silkworm eggs and mulberry seeds out of the country.

9. (   ) Silk production began to spread to Western Europe at the end of the 12th century before the conquests of the Crusades.

10. (   ) Today the technique adopted in silk painting is gutta resists, which is one of the most widely used.

II Outline the "Discover and Spread of Silk Painting" to the world in timeline pattern.

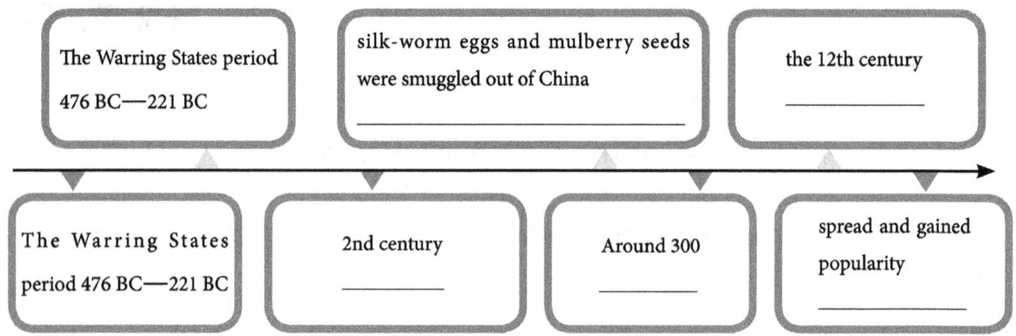

Words and Expressions in Use

I Match the collocations in Column A with the Chinese meanings in Column B.

Column A	Column B
1. be primed to	A. 追溯到……为止
2. date back as far as	B. 易于切割成任何需要的形状
3. reach its height	C. 准备就绪
4. represent the integrity and noble qualities of the man	D. 令人垂涎的商品
5. easy to cut to any desired shape	E. 达到它的鼎盛时期
6. coveted trading commodity	F. 迈向
7. make its way to	G. 代表男人的正直和高贵品质
8. gain popularity in	H. 得到普及

II Compare the following pairs of sentences and explain the different parts of speech and choose the similar meaning of italicized words.

1. Originating from China, silk painting is an art form with over 2,000 years of history that ***involves*** applying coloured pigment to silk cloth.

   (A) Parents should ***involve*** themselves in their child's education.

   (B) These services ***involve*** 11 government departments, and include housing, medical insurance and social security.

(C) His confession ***involved*** a number of other politicians in the affair.

(D) Scientific research often ***involves*** creativity and inventions.

2. Because pigments ***spread*** freely when applied to silk, the artist relies less on brushes and more on creating boundaries for the pigment through the use of a resist.
   (A) Sue ***spread*** the map out on the floor.
   (B) Water began to ***spread*** across the floor.
   (C) Use of computers ***spread*** rapidly during that period.
   (D) Within weeks, his confidence had ***spread*** throughout the team.

3. Silk painting as an art continued to spread and gained ***popularity*** in Britain and America by the 1970s.
   (A) The story had an extensive ***popularity*** among American readers.
   (B) Walking and golf increased in ***popularity*** during the 1980s.
   (C) In politics, power and ***popularity*** are not synonymous.
   (D) People used to call me Mr. ***Popularity*** at high school, but they were being ironic.

4. He is most famous for his paintings of chrysanthemum, and their ***expression*** of traditional Chinese aesthetics.
   (A) ***Expressions*** of sympathy flooded in from all over the country.
   (B) There was a worried ***expression*** on her face.
   (C) This forms the basis for our mathematical ***expression*** for the electric field.
   (D) Laughter is one of the most infectious ***expressions*** of emotion.

5. He ***developed*** a new technique of silk painting, called "Wumo" or "Silent in Suzhou" painting.
   (A) We ***developed*** trade with them for mutual benefit.
   (B) Children with highly ***developed*** problem-solving skills are rare.
   (C) I was sorry he had left, although I soon ***developed*** a productive relationship with his successors.
   (D) The company ***develops*** and markets new software.

6. They brought the art to France where the Serti technique was ***introduced*** in the 1900s.
   (A) The system we ***introduced*** in 1980 has been a great improvement.
   (B) We were ***introduced*** by a mutual friend.

(C) She ***introduced*** me to Sir Tobias and Lady Clarke.

(D) The Government has ***introduced*** a number of other money-saving moves.

Ⅲ Fill in the blanks with the words or expressions given below. Change the form where necessary. Each word or expression can be used only once.

originate from	compare to	be primed to	integrity	alternatively	acclaim
coveted	accommodate	exclusive	a multitude of	is renowned for	smuggle

1. The eatery has been __________ many yakiniku restaurants in town. From its premium selection of meat, you can try a bit of everything because each order starts from one slice of meat.

2. They either____________ Chinese tradition, or are symbols, spiritual meanings or items that have emerged over the course of social development, thus are closely tied to Chinese culture.

3. Because the steam can cook food thoroughly and no turning or stirring is required during the process, the shape and _________ of delicate ingredients can be kept.

4. Luis Suarez has "matured a great deal" and __________ lead Uruguay's assault on the World Cup, coach Oscar Tabarez said on the eve of their opening game in Russia.

5. During an _________ interview last week with Shanghai Daily, he shared some of his thoughts on the rapid development of commercial real estate in China and its prospects going forward.

6. Zheng once covered Taylor Swift's song "Wildest Dreams I Bad Blood" with his friend, which received _______ from netizens and Swift herself.

7. The "Breakthrough Prize" is only six years old but it is far more lavish than the ______ Nobel, which comes with prize money of about US$1 million.

8. I can either send it to you by express mail or_________ compensate you at a reasonable price.

9. This region _____________ tea and silk.

10. Elsewhere, legal shipments of ivory have been used to _______ illegal ivory.

11. Many readers are well aware of the ___________ delicious applications of star anise in

Chinese cuisine, but what about an ideal wine partner?

12. Shelters were built to ____________ those staying in the square outside.

Translation Skills

句子英译 (1)——句子主语的调整

汉语和英语分属不同的语系，差异巨大，尤其是在句法结构和思维模式上。虽然词是汉英两种语言转化活动中最基本的语言单位，但是还不足以表达一个完整的意思，而句子是“由词和词组组成的能表示一个完整意思的话”。汉语的词性和词义也往往需要通过所在的句子来确定。汉译英时，如果只是明确了关键词对应的译文，却没有搭建起正确的句子架构，所呈现的内容就好比一个个割裂的对应的区块，而没有一个完整的版图。

汉语是一种意合的语言，其句法强调逻辑关联和意义关联，句子成分之间多无明示逻辑关系的标记，主语的选择常无词性上的限制，谓语不受主语支配，无人称、数、时态的要求。英语是一种形合的语言，其句法强调形式与功能，句子成分之间的关系要求用形式标记表明，句子的主语只能是由名词或名词性的词语担任，谓语需与主语在人称和数上面保持一致，有时态和语态的变化，这些语法功能如同纽带衔接成一个形式完整而严密的句子。在汉译英时，我们需要牢记两种语言在句法上的差异，同时还要明确一点：无论英语的句式如何变化，都是围绕着几种基本句型而变化的。英语的基本句型为以下几种。

（1）主语（S）+谓语（V）；

（2）主语（S）+谓语（V）+宾语（O）；

（3）主语（S）+谓语（V）+补语（C）；

（4）主语（S）+谓语（V）+间接宾语（O）+直接宾语（O）；

（5）主语（S）+谓语（V）+宾语（O）+补语（C）；

（6）主语（S）+谓语（V）+状语（A）；

（7）主语（S）+谓语（V）+宾语（O）+状语（A）。

以下将会从主语、谓语、状语等成分的确定和调整等方面提供相应的翻译策略，保证在汉译英时成功构句，实现原文到译文的成功转换。

1. 补充主语

汉语句子中，常会出现主语隐含或缺失的情况，这个时候就需要根据语境和英语的句法习惯来补充主语，请看下面几个例句。

（1）近朱者赤，近墨者黑。

原文显然是省略了主语，不难判断这里的主语是指人，而且是泛指的人，因而可以译为One takes the behavior of his company.

（2）必须给子孙后代留下天蓝、地绿、水净的美丽家园。

We should leave to our future generation a beautiful homeland with green fields, clean water and a blue sky.

（3）在西安可以参观秦朝的陵墓和兵马俑。

You can visit the Qin Dynasty Tombs and Terracotta Warriors.

2. 转换主语

上文已经提到汉语句子中的主语并不限于是名词或名词短语，可以是动词、介词短语、副词等，在翻译时，就需要重新确定主语，使译文通顺，符合英文句法，请看以下例句。

（1）要努力制止环境恶化，显著改善环境。

原文并没有明显的主语，但是根据make efforts to这个短语，我们可以把efforts设定成为句子的主语，把译文转化成为被动语态。

Efforts should be made to curb the deterioration of the environment and noticeably improve it.

（2）华北大部分地区因为过于寒冷或过于干燥，无法种植水稻。

按原文的形式看，“华北大部分地区”是主语，但从整个句子的意思来看，这一部分其实是充当了一个地点状语，在译成英文时，真正的主语应该是“水稻”。

Rice cannot grow in the most areas of North China where it is either too cold or too dry.

（3）早在5000年前，中国就开始饲养蚕。

同上句一样，中国看似是主语，其实还是一个地点状语。

Silkworms were domesticated as early as 5,000 years ago.

（4）通过阅读，人们能更好地学会感恩、有责任心和与人合作，而教育的目的正是要培养这些基本素质。

Through reading, people can better learn to be grateful, responsible and cooperative. And these essential qualities are just what education wants to cultivate.

原句分为两个分句，分别是以“人们”和“教育的目的”为主语，但在译成英文时，考虑到前后两个分句的衔接，把原句中的宾语“这些基本素质”调整成为了译文的主语。

（5）众所周知，酗酒有害身体，但适度饮酒是有益的。

这里的众所周知，我们可以套用英文相对应的句型it is known that，即用形式主语it来充当主语。

It is known that excessive drinking is harmful to our health, but moderate drinking is considered beneficial.

（6）近年来，我国国际地位和影响力显著提高。

A. The recent years have witnessed the remarkable increase in China's international prestige and influence.

B. In the recent years, China's international prestige and influence have improved significantly.

比较两句译文，不难发现译文A 将原来的译文“中国的国际地位和影响力”替换成为“近年”，显得更地道。事实上，在正式文体中，常会见到这种以时间作为主语的表达。

3. 沿用原句主语

上文我们举了很多原文和译文主语不一致的例子，但也有不少句子我们是可以把原文的主语作为译文的主语的。例如：

（1）哈佛是最早接受中国留学生的美国大学之一。

Harvard is one of the first American universities to accept Chinese students.

（2）中国剪纸有1500多年的历史，在明朝和清朝时期特别流行。

Paper cutting has a history of more than 1,500 years. It was particularly popular during the Ming and Qing dynasties.

Exercising Your Skills

Identify the subject in the following sentences and translate them into English.

1. 要同国际社会一起积极应对全球气候变化。

____

2. 中国历史上产生了很多杰出的哲学家和思想家。

____

3. 生于忧患，死于安乐。

____

4. 优先发展科技教育。

____

5. 博物馆展出了很多来自亚洲的精美传统服装。

____

Unit Project

In this time of more diverse lifestyle choices for much of the developing world, the desire to dedicate a lifetime to the tedious study of a traditional art or trade is often not very strong, and these traditions are dying out with the masters who will be the last to practice them. Cultures die when their members walk away; traditions go extinct when there is nobody left who wants to practice them. The task to protect our cultural traditions and our arts and crafts lies with the young generations and that is something we need to challenge from now on.

For thousands of years in Chinese history, one of things to judge a girl is her needlework, i.e. her sewing and embroidery skill. As a result, it became a must-have skill for almost every girl and something she learns to do ever since her birth. Although it takes a long time to learn a craft like embroidery, there are still things we can learn to do right away. From **Passage 1**, we have learned that Li Jing is not a born round fan artist and that it is something he picked up later in his life. So there are quite a number of things you can do with some preparation, such as, making a round fan, tying a knot and dyeing a scarf.

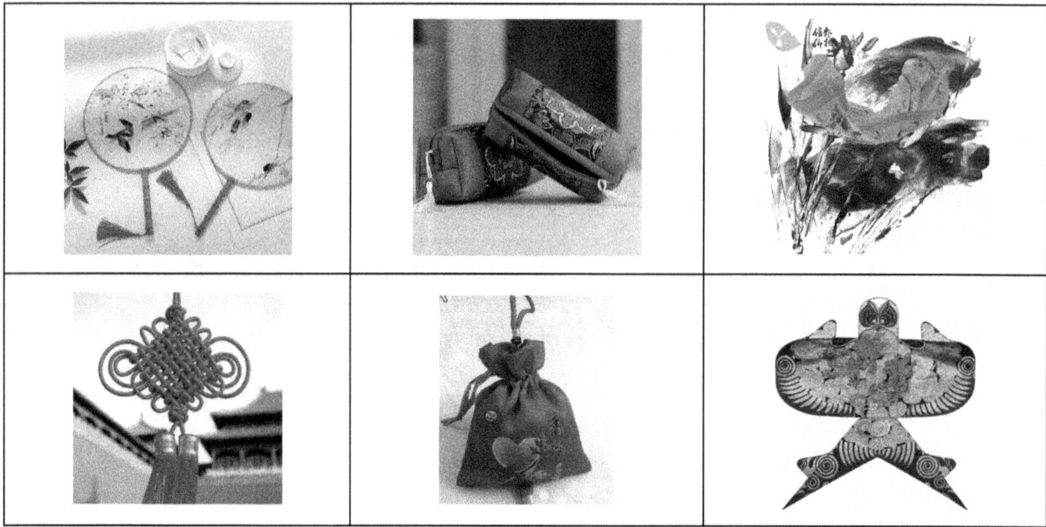

Work in groups to create silk craftwork of your own by using silk materials like silk thread, silk cloth, etc.

Unit 5 Key5

1. Work together to decide what craftwork you are going to produce, what materials to choose and how you are going to go about it.

2. Present to the class your work and tell the class why you choose this project and describe your making process.

Unit 6 Famous Brands

Unit Guide: Brands are the core competitiveness of goods. Silk products are no exception. In China there are many silk brands, some of which, though famous domestically, are still unknown in the international market. This has resulted in a very big constraint for the development of China's silk industry. In **Unit 6**, we will look at efforts being made to transform and upgrade this traditional industry so that silk from China will be restored to its former glamour. As long as we stick to innovations and position our brands according to their true worth, "Made in China" will become a treasured label in the foreseeable future.

Lead-in

I Get to know the brands.

Here are some pictures of brands and names. Choose the correct brand name for each picture, and discuss the following questions.

A. Wensli	B. Dujinsheng	C. Hermès	D. Gucci
E. Marc Rozier	F. Ruifuxiang	G. Xiuniang Silk	

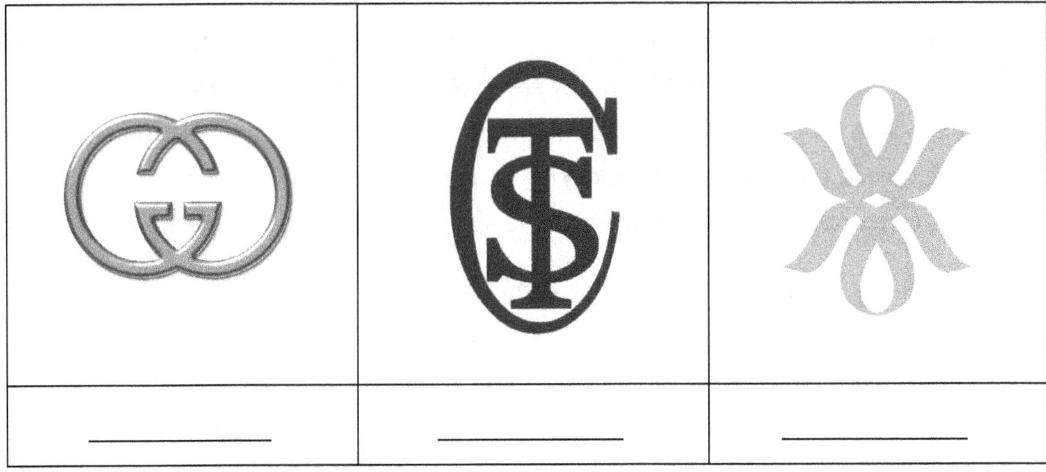

1. Have you ever bought anything with the above brand? Describe your experience.

2. Put those logos of brands in order according to your preference. Why do you put then in that way?

3. Do you know any other silk brands in life? Introduce them in class.

II Group discussion.

1. Read the following case carefully and discuss with your partners.

2. Generalize the Hermes' brand philosophy.

3. Discuss about some inspiration to Chinese silk makers.

Hermes' Brand Philosophy

The brand philosophy of Hermès is deeply entrenched in the platforms of "quality" and "refinement. " According to Hermès, every product coming out under the brand's name should reflect the hard work put into it by the artisan. The main strength of the Hermès brand is the love for craftsmanship. When customer feels the presence of the person who crafted the object, the object is very likely to bring him back to his own sensitivity and pleasure.

With the history of nearly 184 years, the company has passed through multiple generations of the Hermès family, but the principles of the brand have never been diluted. Each Hermès product is entirely manufactured by hand by only one craftsman, which signifies the quality of craftsmanship and uniqueness of its products.

For example, its silk scarves are only made from silk produced by Hermès farms in Brazil.

The driving force behind Hermès is the intense desire to remain exclusive. The aura of exclusivity is important for the company because it only intends to portray the brand and its products as "ultra-premium luxury, " which can be afforded by the very few. In line with these brand philosophies, the company does not have a marketing department. The two core drivers of the company's business engine are intuition and creativity.

Ⅲ Breaking into a butterfly.

视频

Watch the video and fill in the blanks with the exact words you hear.

1. A West Lake panorama drawing has ________________________________ ________. It is not a normal drawing but a top-quality silk knit.

2. Chinese silk once again ________________________through its classic elegance and graceful excellence.

3. A local enterprise Wensli, ________________ had dedicated nearly one year in preparation. Established in 1975, the company has grown up to ________________________.

4. Zhejiang has always ________________________ with its distinct four seasons and fertile soil.

5. A piece of folium with a piece of silkworm cocoon could knit ________________________. Silk knitted __________________________ Chinese historical intelligence, and handcraft may be alienated by times.

6. Traditional skill need to be passed on while ______________________________________. Silk is no longer a simple knitting material but ____________________________________. With the mixture of creative culture, trendy design, high tech, and internet, ____________ ______________________________.

Passage 1

A Very Long and Winding Silk Road

Yan Yiqi

For a product that has been a **staple** of East-West trade for more than 2,000 years, it has an elegant **allure** that never seems to show any signs of age.

Gracing catwalk clothing in the world's fashion capitals and adding a dash of class to **attire** everywhere else, silk **takes some beating**.

Originating in China, it was transported to the rest of the world in large quantities as early as the Western Han Dynasty (206 BC—AD 24). Even today, it is likely that the silk you see anywhere will have come from China.

A report **issued** by China International Silk **Forum**[1], held in November, **reckoned** that over the past decade China made nearly 80 percent of the world's silk. Among the Chinese cities that contribute to that mountain of fabric is Hangzhou, which is said to provide a **quarter** of the global supply.

"In the first half last year, the city exported 44.2 million tons of silk, worth $439 million," the city's customs office says.

"Silk has been a **pillar** industry of Hangzhou for more than 1,300 years, since the Tang Dynasty[2], " says Fei Jianming, chairman of Hangzhou Silk Association. "Hangzhou silk now accounts for 30 percent of the **domestic** market share. There are more than 1,000 silk companies in the city."

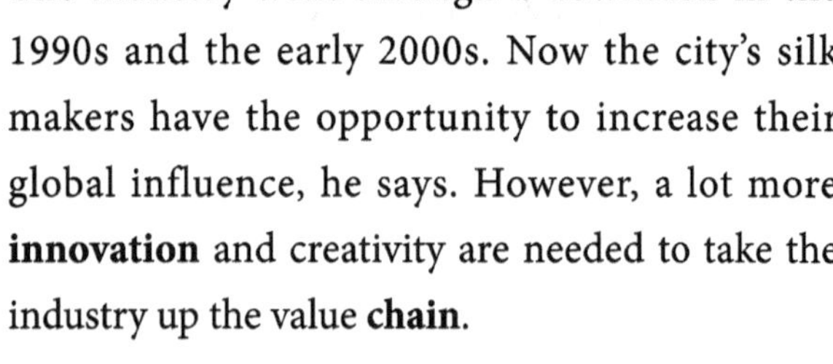

The industry went through a downturn in the 1990s and the early 2000s. Now the city's silk makers have the opportunity to increase their global influence, he says. However, a lot more **innovation** and creativity are needed to take the industry up the value **chain**.

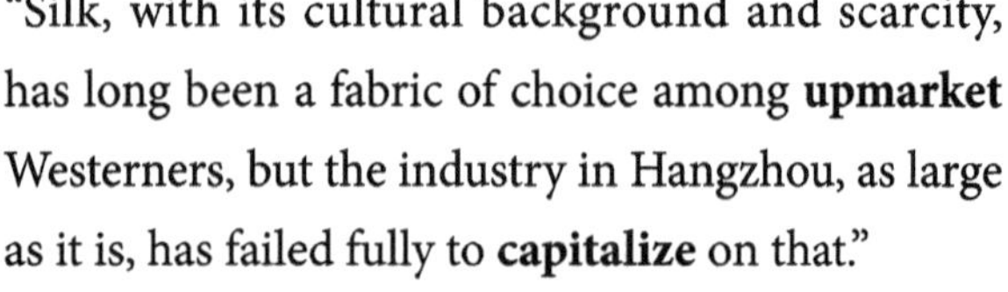

"Silk, with its cultural background and scarcity, has long been a fabric of choice among **upmarket** Westerners, but the industry in Hangzhou, as large as it is, has failed fully to **capitalize** on that."

"In US dollars, the foreign exchange earned through the city's silk exports is only 8 percent that of Italy's and no more than 10 percent of ROK's, " he says.

Tu Hongyan, chairwoman of Wensli Group, the largest silk maker in Hangzhou in terms of sales **revenue**, says that while China produces the vast majority of the world's silk, it does not have a **self-run** silk brand that is known to the world.

"The famous silk brands are in European countries such as France and Italy. Chinese companies have long done OEM (original equipment **manufacturer**) production for those brands. We have done the **bulk** of the work but earned the least money."

"For example, you will pay more than 3,000 yuan ($510) for a **Hermes**-branded silk scarf, but how many people know that that scarf comes from China and that we sell it for 100 yuan or less?"

Wensli has been an OEM for foreign companies for more than 15 years, Tu says **Gucci**, **Prada** and **Dior** are among its customers.

Like Wensli, many of the other silk makers in Hangzhou are OEMs, although many also produce brands that are well-known domestically.

Wu Haiyan, dean of the School of Design of the Chinese Academy of Art[3] in Hangzhou, **echoes** Tu in talking of the **limitation** of the country's silk industry.

"Down the ages, Chinese silk has had a **magnificent reputation**, but **barely** any Chinese silk company appears in the global **showroom**. They are limited by textile techniques, their **inability** to do research and development and their fashion sense. That means they have had to **play second fiddle**, being content to be OEMs for foreign companies."

In the **bleak** period of the 1990s, many silk companies in Hangzhou folded. There were serious questions about whether the industry there could survive, Wu says.

One of those that kept its head above water was Hangzhou Dujinsheng Silk **Weaving Mill**, one of the oldest silk companies in the city. It was established in 1921 and was soon winning widespread acclaim. It was awarded a gold medal at the World Expo[4] in **Philadelphia** in 1926, it says. At its **peak** the company employed more than 3,000 people, with more than 60 designers and 600 weaving machines, says Ke Daocun, the company's chief engineer.

Ke, 72, has **witnessed** the company's many ups and downs in his 40 years of working for it.

"It was one of the largest manufacturing companies in Hangzhou in the 1980s but, since the

1990s, things have gone **downhill**."

"At its lowest point it had just 300 workers and no more than 10 designers—and only 18 machines, " he says.

"They say traditional manufacturing industries will eventually die, but I don't think that time has arrived for silk."

Like Ke, Tu has spent a lot of time **fretting** about the industry in Hangzhou.

"People in the industry have mixed feelings about silk. It is seen as an important cultural symbol and the textile industry looks on it with pride. But there are those who reckon all the **glamour** is something from a **bygone** age. It is high impossible to get that back."

Tu's concerns for the industry are all the more understandable given that she has not only **devoted** her life to silk, but that she follows her mother and four other generations of the family on the same path. Her mother spent the best part of a lifetime trying to sell Hangzhou silk to the world with its own brand, a goal that remained unrealized when she retired in 2012.

For China in recent years, transforming and **upgrading** traditional industries has **posed** a huge challenge. The silk industry in Hangzhou is no exception. Many industry insiders believe one way of doing this is to make more of the possibilities in cultural industries. For example, silk companies in Hangzhou should consider developing exquisite tourism **souvenirs** with silk, Wu says.

"You can find silk souvenir products in markets and in tourism spots anywhere in Hangzhou, but they are not exquisite—and they are far from original. Silk companies need to put a lot more effort into designing these things, in particular giving them a distinct Hangzhou style."

As **downbeat** as some of the **assessments** of the industry's performance by some insiders may be, things are far from grim.

In November, Wensli set up a **strategic cooperation** relationship with **Marc Rozier**, a well-known silk brand in **Lyon**, France.

Marc Rozier will produce silk scarves for Wensli, while Wensli will share its sales channels in China with its French partner.

"The scarves, under our brand, will be designed, dyed and made in France," Tu says. "The price will be two-thirds that of Hermes. "

The scarves will go into the high-end market in China. The company nurses hopes of selling them in global **luxury** markets such as France and Italy, Tu says. "The strong impression that this **collaboration** has left me with is that silk making in France is highly valued. They pay extreme attention to the industry and I think that is why 'Made in France' is such a **treasured label**."

Tu says Wensli's goal is to become China's Hermes in the luxury market and to keep pace with global fashion.

It **augurs** well for silk in Hangzhou, Fei says.

"Hangzhou has its advantages in developing the silk industry, including its cultural background, the industrial chain that is already in place and the **diversified** competition. Not every company should necessarily do what Wensli is doing. But so long as it sticks to innovation and **positions** its brands accordingly, I think the future is bright."

Words

staple ['steɪpl] *n.* 主要产品；大宗物品
allure [ə'ljʊə(r)] *n.* 诱惑力；魅力
grace [greɪs] *vt.* 使优美
attire [ə'taɪə] *vt.* 打扮；使穿衣
n. 服装；盛装
beating ['biːtɪŋ] *n.* 打；敲打
originate [ə'rɪdʒɪneɪt] *vi.* 发源；发生
issue ['ɪsjuː] *vt.* 发行；发布
forum ['fɔːrəm] *n.* 论坛；讨论会
reckon ['rekən] *vi.* 估计；计算
quarter ['kwɔːtə] *n.* 四分之一
pillar ['pɪlə] *n.* 支柱；核心
domestic [də'mestɪk] *adj.* 国内的；家庭的
innovation [ˌɪnə'veɪʃn] *n.* 创新；革新
chain [tʃeɪn] *n.* 链
upmarket ['ʌpˌmɑːkɪt] *adj.* 高端的；高级的

capitalize [ˈkæpɪtəlaɪz] *vt.* 使资本化
revenue [ˈrevənjuː] *n.* 税收收入；财政收入
self-run [selfˈrʌn] *n.* 自营
manufacturer [ˌmænjʊˈfæktʃərə(r)] *n.* 制造商；厂商
bulk [bʌlk] *n.* 大多数；大部分
echo [ˈekəʊ] *vi.* 随声附和；发出回声
limitation [lɪmɪˈteɪʃn] *n.* 限制；限度
magnificent [mægˈnɪfɪsnt] *adj.* 壮观的；杰出的；精彩的
reputation [repjʊˈteɪʃ(ə)n] *n.* 名声；声望
barely [ˈbeəli] *adv.* 几乎不；仅仅
showroom [ˈʃəʊruːm] *n.* 陈列室；样品间
inability [ˌɪnəˈbɪləti] *n.* 无能力；无才能
fiddle [ˈfɪdl] *n.* 小提琴
bleak [bliːk] *adj.* 黯淡的；无希望的
weaving [ˈwiːvɪŋ] *n.* 织动；编织
mill [mɪl] *n.* 工厂；磨坊
peak [piːk] *n.* 顶峰；顶点
witness [wɪtnɪs] *vt.* 目击；证明
downhill [daʊnˈhɪl] *adv.* 向下；每况愈下
fret [fret] *vi.* 烦恼；焦急
glamour [ˈglæmə(r)] *n.* 魅力；诱惑力
bygone [ˈbaɪgɒn] *adj.* 过去的
devote [dɪˈvəʊt] *vt.* 致力于；奉献
upgrade [ʌpˈgreɪd] *vt.* 使升级；提升
pose [pəʊz] *vt.* 造成；形成
souvenir [ˌsuːvəˈnɪə(r)] *n.* 纪念品；赠品
downbeat [ˈdaʊnbiːt] *adj.* 悲观的；不强烈的
assessment [əˈsesmənt] *n.* 评定；估价
strategic [strəˈtiːdʒɪk] *adj.* 战略上的；战略的
cooperation [kəʊˌɒpəˈreɪʃn] *n.* 合作；协作
luxury [ˈlʌkʃəri] *n.* 奢侈品；奢华
collaboration [kəlæbəˈreɪʃn] *n.* 合作；勾结
treasured [ˈtreʒəd] *adj.* 珍贵的；宝贵的
label [ˈleɪbl] *n.* 标签；商标

augur [ˈɔːɡə(r)] *vi.* 预言；预示
diversified [daɪˈvɜːsɪfaɪd] *adj.* 多样化的；各种的
position [pəˈzɪʃn] *vt.* 定位；把……放在适当位置

Useful Expressions

take some beating 独领风骚
play second fiddle 居次位；当副手

Proper Names

Hermes 爱马仕（Hermès），世界著名奢侈品品牌
Gucci 古驰，意大利时尚名牌
Prada 普拉达，意大利时尚名牌
Dior 迪奥，法国奢侈品牌
Philadelphia （地名）费城（美国宾夕法尼亚州东南部港市）
Marc Rozier 马克・罗茜，法国里昂一家有着120年历史的丝绸企业
Lyon （地名）里昂（法国城市）

Notes

1. China International Silk Forum is sponsored by the China Silk Association. In 2015, it was held in Hangzhou, and the theme of the 2015 China International Silk Forum was "silk and fashion."
中国国际丝绸论坛作为世界丝绸行业最高层次论坛，是由中国丝绸协会发起举办的。2015年的中国国际丝绸论坛在杭州举行，其主题是"丝绸与时尚"。

2.Tang Dynasty is the very important period for the spread of silk products and its culture. About 1,300 years ago, when Princess Wencheng of the Tang Dynasty left the capital Chang'an (present-day Xi'an in Shanxi Province) for the Tubo Kingdom, She brought many things as her dowry for the marriage with Songtsan Gambo. Among those treasures were the technique of silk production and embroidery.
唐代是丝绸产品及其文化传播的重要时期。大约1300年前，当唐代文成公主离开首都长安（今陕西省西安市）去吐蕃王国时，她随带了很多东西作为与松赞干布结婚的嫁

妆，其中包括丝绸生产和刺绣技术。

3. The Chinese Academy of Art China, also known as China National Academy of Fine Arts, is a fine arts college jointly constructed by the Ministry of Education, the Ministry of Culture and the People's Government of Zhejiang Province. It was founded in Hangzhou by Cai Yuanpei in 1928. It was the first art university and first graduate school in Chinese history. The academy has many renowned artists in its alumni, such as Lin Fengmian, Pan Tianshou, Huang Binhong, Wu Guanzhong, and is considered one of the most prestigious art institutions in the country.
中国美术学院前身是国立艺术院，于1928年由蔡元培先生发起，在杭州成立，是国内历史最悠久、学科最完备、办学规模最齐整的第一所综合性国立高等艺术院校，以及第一所具有本科和研究生学历教育的高等艺术学府。几十年来，中国美术学院聚集和培养了众多艺术名家，如林风眠、潘天寿、黄宾虹、吴冠中等等，它是国内公认的最负盛名的艺术院校之一。

4. The World Expo firstly held in London, England in 1851, is a large international exhibition designed to showcase achievements of nations. These exhibitions vary in character and are held in different parts of the world. World Expos encompass universal themes that affect the full gamut of human experience, and they are held every 5 years.
1851年英国伦敦举办了第一届世界博览会。此后，世界博览会因其发展迅速而享有“经济、科技、文化领域内的奥林匹克盛会”的美誉。世界博览会每隔五年举行一次，涵盖了影响人类全部体验的普遍主题。

Reading Comprehension

Choose the best answer to each of the following questions or senterces.

1. According to this passage, how long has the silk been a staple product between the eastern and western countries?
   (A) Less than 2,000 years. (B) More than 2,000 years.
   (C) Less than 1,500 years. (D) Almost 1,500 years.

2. When did the silk in large quantities begin to be transported to the rest of the world?
   (A) During the Western Zhou Dynasty. (B) During the Qin Dynasty.
   (C) As early as the Ming Dynasty. (D) As early as the Western Han Dynasty.

3. With over 1,000 silk companies in the city, Hangzhou silk now accounts for________________.
   (A) 40 percent of the domestic market share.

(B) 40 percent of the global market share.
(C) 30 percent of the domestic market share.
(D) 30 percent of the global market share.

4. If a Chinese company did OEM production, how much will it be paid for a silk scarf by Hermes?
   (A) 200 yuan or less. (B) 100 yuan or less.
   (C) 3,000 yuan. (D) 1,000 yuan.

5. Chinese silk company seldom appears in the global showroom. Which of the following reasons is NOT true?
   (A) Chinese silk has had a good reputation.
   (B) Chinese silk companies are limited by textile techniques.
   (C) Chinese silk companies are not capable enough to do research and development.
   (D) Chinese silk companies have not enough fashion sense.

6. Which of the following statements is NOT true about Dujinsheng Silk Weaving Mill?
   (A) It was one of the oldest silk companies in Hangzhou.
   (B) It was awarded a gold medal at the World Expo.
   (C) The company hired more than 3,500 people at its peak.
   (D) Since the 1990s, the company has gone downhill.

7. In recent years, ________________ traditional industries has posed a huge challenge for China.
   (A) transforming and upgrading (B) transforming and expanding
   (C) reforming and upgrading (D) reforming and expanding

8. Which of the following was NOT included in the cooperation between Wensli and Marc Rozier?
   (A) Marc Rozier will produce silk scarves for Wensli.
   (B) Wensli will share its sales channels in China.
   (C) The scarves will be designed, dyed and made in France.
   (D) The scarves will be under the brand of Marc Rozier.

Summary

The following is a summary of Passage 1. Fill in the blanks with appropriate words.

Silk trade has been lasting for more than two thousand years. 1.________ in China, it was always a popular product among the westerners. Over the past 2.________ China produced almost 80 percent of the world silk. As a 3.________ industry for more than 1,300 years, silk made in Hangzhou has made up a quarter of the global supply. Although silk makers now in the city have the 4.________ to increase their global influence, a lot more 5.________ and creativity are needed to take the industry up the value chain. Silk makers in Hangzhou, including Wensli, produce famous brands 6.________, but many of them are OEMs. Limited by textile 7.________ and lack of fashion sense, Chinese silk company 8.________ appears in the global showroom. It is urgent to 9.________ and upgrade traditional industries. To meet the needs of market, silk companies not only should consider developing 10.________ silk products, but start 11.________ with foreign companies. Therefore, the silk products will go into the 12.________ market in China. Most importantly, it makes it possible to render Chinese companies to keep pace with global fashion. As long as silk companies stick to innovation and 13.________ its brands accordingly, the future must be bright.

Word Building

I The prefix *up* can be used together with other words to form new words, meaning "higher / better; upward; toward the top or north of something." Match the following words with their parts of speech and Chinese meanings.

	upload	在舞台后方
n.	upmost	根除
	upwards	向上翘的
adv.	upstage	优势
	uplift	最重要的
adj.	uproot	向上推
	upside	向上
v.	upturned	升高一级
	upscale	上传
	upthrust	抬起

II The suffix *ful* can be used together with other words to form new words, meaning "full of" or "number or quantity that fills or will fill." Match the following words with their Chinese meanings.

	mouthful	满箱
	cupful	一屋子
	bagful	满口
	roomful	一把
	spoonful	满杯
n.	armful	一袋子
	tankful	满桶
	houseful	一抱的量
	handful	一匙
	plateful	满屋
	bucketful	一满盘

Words and Expressions in Use

I Fill in the blanks with the words or expressions given below. Change the form where necessary. Each word or expression can be used only once.

position	collaboration	take some beating	reckon	upgrade	account for
echo	a quarter of	innovation	assessment	showroom	staple

1. Even if you manage to find a bargain, seasoned gemstone collectors ________ that you may need to hold the stones for as long as ten years to get a decent return.

2. Shocking statistics show one-third of workplace accidents occur in the agriculture sector which ____________ just 7.5% of the workforce.

3. An alliance allows its partners to speed up the processes of __________ and market expansion.

4. There was no proper risk __________ done for that kind of industrialized farming.

5. The Chinese Fine Arts Society, in ____________ with the Art Institute of Chicago, will organize a "Lantern Procession" through Millennium Park to Maggie Daley Park,

followed by a lion dance, art making and ice skating.

6. They suspect staff of advising customers to make false claims to replace and _______ old phones.

7. Rapid Deployment, like many of the horse trained by Hughes, has not been at the top of his game but Hughes is on the way back and his recent record in this event _______________.

8. The ________ of the trade between East and West were tropical goods impossible to be produced in temperate Europe—pepper, silk and tea from China, coffee from Java, cotton from India.

9. This opinion was ________ by a few other visitors, who also felt that the prices were competitive.

10. This time, larger __________ and stores arm themselves with more sales persons to manage the huge crowds.

11.Wang firmly __________ Beijing as a still-steadfast proponent of economic globalization.

12. They have no choice but to turn in for work the week after half term knowing that ____________ their classes are still enjoying themselves on holiday somewhere around the world.

II Render the following Chinese expressions into English. The first letter of each word has already been given.

1. 传统制造业 t__________ m__________ i__________

2. 全球四分之一供应量 a __________ of the g________ s__________

3. 国内市场份额 d__________ m__________ s__________

4. 东西方贸易 E__________-W__________ t__________

5. 外汇 f__________ e__________

6. 高端市场 h__________ m__________

7. 增强全球影响力 i__________ the g__________ i__________

8. 研发丝绸旅游纪念品 d________ t__________ s__________with s________

9. 为其他品牌贴牌生产 d_________ O_________ p_________ for o_________ b_______

10. 分享销售渠道 s__________ the s_________ c_________

Translation

Translate the following paragraph into English.

杭州以生产包括绸缎在内的优质丝绸而闻名。分层编织工艺耗费巨大劳力，生产出来的奢华面料有紧贴肌肤之感，令人赞叹。纵观其悠久的历史，杭州各种丝绸和绸缎销往世界各地。购买丝绸的最佳地点是丝绸城，它是全国最大的丝绸批发和零售市场，拥有600多家丝绸企业，经营各种纯真丝面料、服装、工艺品、围巾和领带。那里的商店不仅在国内开展业务，还将丝绸产品出口到其他国家。市场以提供标准化的管理、可靠的质量和公道的价格为其显著特色。

Passage 2

When Lao, Jiao, Fu Rose to Become China's Top Silk Brand

Qiao Zhengyue

The former home of the **upper-class** Laou Kai Fook silk **emporium** on Nanjing Road E., now housing the American fashion brand Forever 21, has just completed a **restoration** of the **facade**.

The historical building at 257 Nanjing Road E. **was** once **home to** an upper-class silk emporium, the Laou Kai Fook & Co Ltd, in the 1930s. It has become a **flagship** store of American **cheap-chic retailer**, which just completed a restoration of the facade.

According to Tongji University Professor Qian Zonghao, author of *Nanking Road 1840s—1950s*, Laou Kai Fook was **originally** named Kai Fook and was established in 1860 on Jiujiang Road by the Zhu brothers, who were from Fujian Province. They later sold the shop to Cheng Yongliu of Suzhou.

The shop's name was later changed to "Laou Kai Fook"—from the **nicknames** of Cheng and his new partners.

Bearded Li Shanshi was called Lao Shou Xing, or "Old God of Longevity[1]." Tall, slim Yao Yingsheng was nicknamed Chang Jiao, or "Long Leg," while Cheng himself—short and fat—was named Da A Fu, or "Big Blessing[2]." They picked one word from each nickname, "Lao," "Jiao" and "Fu" to become "Laou Kai Fook."

The store later moved to No. 257 Nanjing Road E. and grew to become the city's largest silk shop, which not only sold top-quality silk products, but also designed popular patterns. The shop came to be known as "The King of Chinese Silk." It also sold wool fabrics and offered **tailor-made** services.

Professor Qian said the building **came up** following the **widening** of Nanjing and Henan roads.

A news story in *The China Press*[3] in 1935 was the reason for his **assumption**. The report says the destruction of the old buildings on the **property** would allow the widening of Nanjing and Henan roads **in accordance with** the policy of the Shanghai Municipal Council. The facade of the building had a broad **curve** at the corner, just like many of the newer buildings back then.

The report also gave a description of the six-story building that was designed as S. A. **Hardoon** Building for the Hardoon **Estate** by architect Percey Tilley.

"The building will have black **marble** on the first floor, while the upper stories will be of brick. It will contain stores, offices and small apartments," it said.

Laou Kai Fook turned the shop into a **paradise** of silk goods. In 1938, *The China Press* sent a reporter to write about the shop whose eyes were **struck upon entry** into the silk store at 257 Nanjing Road E. by "a 40-inch-wide crepe remained, completed served with the brightest design of **multi-coloured**, tiny flowers."

"But that impression is soon lost, and one is left **bewildered** by the endless variety of silk and satin, flowered crepes, the height of fashion this year, are available by the **tableful**," the reporter described.

The reporter noted the difficulty of choosing among so many attractive materials. However, the shop's service was remarkable, whose staff "requests clients to be at their leisure—no one minds if a customer remains choosing and constantly changing orders for hours **on end**."

"**Apart from** the flowered things, there are delicate **matt** white silks, which would make up into the most **dignified** gowns for formal evening wear—or striped silks in colours or cool blacks and whites, these latter for wise women who recognize the immense **certitude** and sense which a coolly reserved black and white **ensemble** gives to **maturity**. Crinkled crepes, kindly creatures who wouldn't know the use of an iron if they saw one, suggest prettily that they would like to accompany you on your travels, or point out that you can look as fresh at the end of the day as at the start. In short, say 'Everything in Silks' and you say Laou Kai Fook," the reporter wrote.

Another Tongji University Professor Chang Qing added that Laou Kai Fook & Co had made textile goods for the **Cathay Hotel**, including curtains, **slipcovers**, bed sheets and other textile decorations. All these exquisite textile goods were made from high-quality silk of Suzhou within half a year. Even famous American actor **Charlie Chaplin** purchased dozens of silk shirts from the shop when he visited Shanghai in the 1930s.

"The building has become one of the **landmarks** as part of Shanghai's urban memory, along with the **Sassoon House** and **Whiteway Laidlaw** Building on Nanjing Road," said Professor Chang, author of the book *Nanjing Road—Origin of a Metropolis.*

"Laou Kai Fook and Hope Brothers which sold clocks and watches were two traditional brands on Nanjing Road. One Chinese and one Western, they have **merged** into daily life of Shanghai people for a very long time," he said.

While Laou Kai Fook was having good business at 257 Nanjing Road E., another century-old company, **Siemens** China, opened a new office and showroom on the fourth floor of the same Nanjing Road **edifice**.

According to *The China Press* in 1937, the offices had been completely **decorated** and **furnished** in modern style, offering a pleasing note in business quarters. The showroom on the ground floor adjoining the building entrance displayed the company's varied wares in a **striking** setting.

"The **colour scheme** of the decoration is sea green, ivory and black, with counters and desks of simple, modern style. The visitors' **attention is drawn** to a white colored frame extending the entire length of one wall, containing artistically **executed** advertisements for lighting fixtures, irons, electric stoves and other **appliances**, while **quartz** lamps, meters, wiring devices and countless other items are tastefully displayed throughout the room," the report described.

In the 1930s, Siemens China was a branch of the German Siemens Works and maintained offices in today's Guangzhou, Wuhan, Hong Kong, Nanjing, Beijing and Tianjin.

After 1949, Laou Kai Fook became a state-owned company but its fate on Nanjing Road has changed during the times. The building was renovated twice in 1993 and 1999 and later faced a big renovation when Huangpu District was upgrading business in the eastern section of Nanjing Road.

According to a report on *Textile Decoration & Technology*[4] in 2012, to house young customers of the east section of Nanjing Road, Laou Kai Fook **retreated** to smaller shops in communities and maintained only a tailor-made studio in the former department store which had been renovated to be a big flagship of American fashion retailer Forever 21 Inc.

Famous for **churning out** new styles quickly, the **Los Angeles** clothing maker joined other foreign fashion brands pursuing growing Chinese retail market.

The four-story, 8,000-square-meter **branch** in the 1930s silk emporium was one of its largest flagships around the world and was regarded as a symbol that middle and small American enterprises entering the attractive Chinese market of expected 400 million middle-class consumers in 2020.

The silk paradise of the 19th-century Laou Kai Fook has **vanished** from Nanjing Road, but a new Chinese silk brand, Xiuniang Silk from Suzhou which has new concept of design on traditional silk products, restored the 1930s British Whiteaway, Laidlaw & Co Ltd building and opened a grand flagship on the **Bund** end of Nanjing Road.

The **dramatic** change of **century-old shops** on Nanjing Road is a mirror of Chinese brands, old or new, **fronting** the change and **challenges** of times.

Words

upper-class [ˈʌpəˈklɑːs] *adj.* 上流社会的；上层阶级的

emporium [emˈpɔːrɪəm] *n.* 商场；商业中心

restoration [restəˈreɪʃn] *n.* 恢复；修复

facade [fəˈsɑːd] *n.* 外表；建筑物的正面

flagship [ˈflæɡʃɪp] *n.* 旗舰；同类中最成功的商品

cheap-chic [tʃiːpˈʃiːk] *n.* 廉价时尚；便宜服饰

retailer [ˈriːteɪlə] *n.* 零售商；批发商

originally [əˈrɪdʒɪnli] *adv.* 最初，起初；本来

nickname [ˈnɪkneɪm] *n.* 绰号；外号

tailor-made [ˈteiləˈmeɪd] *adj.* 特制的；裁缝制的

widening [ˈwaɪdnɪŋ] *n.* 拓宽；加宽

assumption [əˈsʌmpʃn] *n.* 假定；猜想

property [ˈprɒpəti] *n.* 财产；所有权

curve [kɜːv] *n.* 曲线；弯曲

estate [ɪˈsteɪt] *n.* 房地产；财产

marble [ˈmɑːbl] *n.* 大理石；大理石制品

paradise [ˈpærədaɪs] *n.* 天堂

entry [ˈentri] *n.* 进入；入口

multi-colored [ˈmʌltiˈkʌləd] *adj.* 多色的

bewildered [bɪˈwɪldəd] *adj.* 感到眼花缭乱的；眩目的

tableful [ˈteɪblfʊl] *n.* 一桌东西；一桌人

matt [mæt] *adj.* 无光泽的；不光滑的

dignified [ˈdɪgnɪfaɪd] *adj.* 有气质的；高贵的

certitude [ˈsɜːtɪtjuːd] *n.* 确信；坚信

ensemble [ɒnˈsɒmbl] *n.* 全套服装；总效果

maturity [məˈtʃʊərəti] *n.* 成熟；完备

slipcover [ˈslɪpˌkʌvə(r)] *n.* 沙发套；套子；覆盖物

landmark [ˈlændmɑːk] *n.* 地标；里程碑

merge [mɜːdʒ] *vi.* 合并；融合

edifice [ˈedɪfɪs] *n.* 大厦；大建筑物

decorate [ˈdekəreɪt] *vt.* 装饰；布置

furnish [ˈfɜːnɪʃ] *vt.* 布置；装备

striking [ˈstraɪkɪŋ] *adj.* 显著的，突出的

scheme [skiːm] *n.* 方案；组合

execute [ˈeksɪkjuːt] *vt.* 实行；执行

appliance [əˈplaɪəns] *n.* 器具；器械；装置

quartz [kwɔːts] *n.* 石英

retreat [rɪˈtriːt] *vi.* 撤退；后退

churn [tʃɜːn] *vi.* 搅拌；搅动

branch [brɑːntʃ] *n.* 分支；分部

vanish [ˈvænɪʃ] *vi.* 消失；突然不见

Bund [bʌnd] *n.* 外滩（上海）；东亚各国的堤岸

dramatic [drəˈmætɪk] *adj.* 巨大的；令人吃惊的

front [frʌnt] *vt.* 面对；朝向

challenge [ˈtʃælɪndʒ] *n.* 挑战

Useful Expressions

be home to　是……之乡；是……所在地
come up　出现；开始
in accordance with　依照；与……一致
strike upon　打动；撞击
on end　连续地；竖着
apart from　除……之外；且不说
colour scheme　色彩设计
draw attention to　吸引对……的注意力
churn out　大量生产
century-old shops　百年老店

Proper Names

Lao, Jiao, Fu　老介福，1860年由福建祝氏两兄弟合资创办，经营高档绸缎。文中又称作 Laou Kai Fook
Hardoon　哈敦（人名）
Cathay Hotel　国泰航空酒店
Charlie Chaplin　查理・卓别林（英国电影演员，导演，制片人）
Sassoon House　沙逊大厦，1926年兴建，得名于当时著名英国商人维克多・沙逊爵士
Whiteway Laidlaw　怀特威・莱德劳
Siemens　德国西门子公司，是德国的跨国企业
Los Angeles　（地名）洛杉矶（美国城市）

Notes

1. Old God of Longevity, also known as the star of Shou, who lives in the star of the south pole in Chinese astronomy, and is believed to control the life spans of mortals. According to legend, he was carried in his mother's womb for ten years before being born, and was already an old man when delivered. He is recognized by his high, domed forehead and the peach which he carries as a symbol of immortality. The longevity god is usually shown smiling and friendly, and he may sometimes be carrying a gourd filled with the elixir of life.

老寿神，也被称为寿星。在民间，人们认为供奉这位仙神，可以健康长寿。这其实是道教追求长生的一种信仰。传说寿星出生前，在母亲的子宫里呆了很多年，到出生时已经是一位老人。寿星高高突出的圆额头和手持的桃子，被当作长寿的象征。寿星通常面带微笑友而且善，有时他也随身带着一个充满生命灵药的葫芦。祭祀寿星的活动在东汉时候就有记载，当时与敬老仪式相结合，人们要向七十岁左右的老人赠送拐杖。

2. Big Blessing, or Da A Fu, a plump boy holding a fish or a qilin (a mythical animal believed to punish the wicked), is the most popular figure in the clay art. The legend goes that long, long ago, Huishan was plagued by a ferocious lion that devoured children, and the people prayed for help to get rid of the lion. One day, a god, disguised as a boy, fought the beast and finally subdued it. To express appreciation and to commemorate the god's beneficence, local people started to make clay figures of their boy saviour Da A Fu.

大阿福，一个拿着鱼或麒麟（一种被认为是惩罚邪恶的神话动物）的胖男孩，是粘土艺术中最受欢迎的人物。传说很久以前，惠山被一只吞噬孩子的凶猛狮子所困扰，人们祈求上苍帮助摆脱狮子。有一天，一个伪装成男孩的上神，与野兽搏斗并最终制服它。为了表达感激并且纪念上神的仁慈，当地人开始用粘土塑造救星大阿福的形象。

3. *The China Press* founded in New York in 5 January, 1990, is a Chinese-language newspaper published in the United States. Statistics show that *The China Press*, along with *Ming Pao*, *Sing Tao*, and the *World Journal*, are the major newspaper of overseas Chinese community in the United States and in Canada.

《侨报》于1990年1月5日在纽约创刊，是一份在美国出版的中文报纸。《侨报》报道风格较为严肃，传播内容以新闻、时事报道、专题为主，与《明报》《星岛日报》《世界日报》等是美国和加拿大海外华人社区的主要报纸。

4. *Textile Decoration & Technology* is a professional journal in Chinese textile industry. With a history of thirty more years, the journal has played an important role in upgrading the textile technology, promoting exchange of the relative information and enhancing the development of textile industry.

《纺织装饰与技术》，中国纺织工业专业期刊。该期刊已有三十多年的历史，在提升纺织技术、促进相关信息交流、促进纺织业发展方面发挥了重要作用。

Reading Comprehension

Decide whether the following statements are true or false. Write "T" for true and "F" for false.

1. (　　) Laou Kai Fook was established in 1860 on Nanjing Road by the Zhu brothers.

2. (　　) The name "Lao Jiao Fu" derived from the nicknames of Cheng and his new partners.

3. (　　) "Jiao" in the shop's name originally means "Big Blessing."

4. (　　) Known as " The King of Chinese Silk, " Laou Kai Fook sold only silk products in the past.

5. (　　) It was quite often that a customer of the shop could change order constantly for hours.

6. (　　) Famous Australian actor Charlie Chaplin bought many silk shirts from the shop during his visit to Shanghai.

7. (　　) According to The China Press in 1937, the offices of the shop had been decorated in classic old style.

8. (　　) In the 1930s, Siemens China maintained its office in many Chinese cities, including Nanjing.

9. (　　) By the year of 2012, Laou Kai Fook completely disappeared from the Nanjing Road in Shanghai.

10. (　　) Xiuniang Silk from Suzhou opened a grand flagship on the Bund end of Nanjing Road.

Words and Expressions in Use

I Match the collocations in Column A with the Chinese meanings in Column B.

Column A	Column B
1. Shanghai's urban memory	A. 一楼展示厅
2. middle-class consumers	B. 日益增长的中国零售市场
3. the showroom on the ground floor	C. 上海城市记忆
4. endless variety of silk and satin	D. 国有企业
5. growing Chinese retail market	E. 无数种丝绸锦缎
6. top-quality silk products	F. 中国品牌的写照
7. a state-owned company	G. 质量顶级的丝绸产品
8. a mirror of Chinese brands	H. 中产消费者

II Compare the following pairs of sentences and choose the one with the same meaning as that in the given sentence.

1. ...quartz lamps, ***meters***, wiring devices and countless other items are tastefully displayed.
   (A) He placed third in the 1,000 ***meters***.
   (B) Record the number that the ***meter*** reads.
   (C) Up till now, the horticulturalists' water use has not been ***metered***.
   (D) Jonathan Van ***Meter***, a writer at New York magazine, blames fashion and celebrity glossies filled with images of teenagers for the New New aesthetic.

2. ... to ***house*** young customers, Laou Kai Fook retreated to smaller shops in communities.
   (A) The English and Scottish royal ***houses*** had become closely connected through marriage.
   (B) I couldn't keep my room clean and barely knew how to take care of two ***house*** cats.
   (C) The Reptile ***House*** gives visitors an insight into the reptile and amphibian world in its many shapes and sizes.
   (D) More than 500 students and nurses will be ***housed*** in a huge new accommodation block situated in the heart of the city centre.

3. ... the shop's service was remarkable, whose staff requests clients to be at their ***leisure***...
   (A) For one thing, you have the time and opportunity to read it and think about it in total freedom and at ***leisure***.
   (B) Sport and ***leisure*** activities are the main focus today.
   (C) This enabled people to wander through at their ***leisure*** and view the school as a

whole.

(D) On the other hand, I think I'd be a really good lady of ***leisure***.

4. ...no one ***minds*** if a customer remains choosing and constantly changing orders...
   (A) "You have to stop ***minding*** being teased," I said, "And no one will tease you."
   (B) We have to turn our ***minds*** and attention to the serious challenge about what to do about social conditions.
   (C) She fought her ***mind*** and was determined that her husband was a good and caring man.
   (D) However, I believe nightmares are a gift of our subconscious to our conscious ***minds***.

5. ...creatures wouldn't know the use of an ***iron*** if they saw one...
   (A) A magnet is the device that attracts certain types of metals, like ***iron*** or steel.
   (B) They suggest that recovering patients reduce ***iron*** in their diet.
   (C) Her father had a will of ***iron***.
   (D) Smaller but still significant numbers of people buy electric ***irons*** and kitchen equipment.

6. ...Laou Kai Fook had made textile goods for the Cathay Hotel, including bed ***sheets***...
   (A) Four hours later I woke up, my eyes fluttering open, the ***sheets*** around me damp with sweat.
   (B) Just as she closed her eyes again a huge gust of wind smashed her windows open, sending ***sheets*** of paper from her desk to the floor.
   (C) Return the answer ***sheet*** printed at the end of the article and fill out all sections carefully.
   (D) It will freeze like a ***sheet*** of ice on your face the minute you go out.

III Fill in the blanks with the words or expressions given below. Change the form where necessary. Each word or expression can be used only once.

assumption	merge	come up	vanish	apart from	tailor-made
striking	restoration	bewildered	exquisite	landmark	design

1. And even though the design is quite ________on its own, the table doesn't take up a lot of visual space, and therefore allows the modern shape to seamlessly blend with more traditional aesthetics.

2. How so? If Chinese do not even do the Chinese drawing, this kind of art will gradually

_______ from the stage of history.

3. Many brands and types of products are specifically _________ to protect and restore your leather goods.

4. Many, although as we saw earlier not necessarily all, clients get some benefit from ___________ nursing care.

5. I couldn't begin to explain my enthusiasm (much less the style and content of the films) to the __________ friend who accompanied me.

6. There was a(an) __________ that a lot of those paintings would come to the museum, notably the Hopper.

7. So we keep track of types of questions that ________, unless I said something stupid in class or my words were confusing, certainly.

8. We try our best to make Chinese culture ______ into the world's cultural, economic and social development, to make it meet the need of the time.

9. ________ the fragrant wintersweet, rivers in the garden are clearer than before and part of its beams, walls and grounds have been repaired, offering visitors with a safe and pleasant touring environment.

10. One of the country's historic attractions is Mingtepa city in the Andizhan region, a crucial ________ on the ancient Silk Road providing valuable clues about the ancient Da Yuan kingdom.

11. The sight of the centuries-old structure, covered in ornate mosaics and undergoing _________, struck the young artist with awe.

12. With beautiful scenery and ________ costumes, Coppelia is a truly delightful production to charm audiences of all ages.

Translation Skills

句子英译（2）——谓语的选择

汉译英时，谓语的选择需要遵循以下原则：

（1）与主语在人称和数上保持一致；

（2）谓语时态符合原句意义；

（3）与主语搭配一致；

（4）与宾语搭配一致；

（5）若原文是连动式谓语，译文有且只有一个动词作为谓语，其他动词采用其他形式。

1. 人称和数的主谓一致

廉租房虽然帮到买不起房子的人，但是会增加人们对政府的依赖。

Although state housing benefits those who cannot afford their own house, it can increase citizens' dependence on the government.

这里的housing请注意是个单数的概念，所搭配的谓语需是第三人称单数。

2. 时态符合原句意义

自从他出国后，我从未听到过他的消息。

I have never heard about him since he went abroad.

原文中的"自从"和"从未"很清楚地提示本句需采用现在完成时，翻译时首先应避免一些低级的语法错误。

3. 与主语搭配一致

随着全球化进程的推进，英文流利的人有更多工作机会。

A. As globalization develops, people who can speak English fluently have more job opportunities.

B. As globalization proceeds, there will be more job opportunities available for those who are fluent in English.

两句译文相比，高下立现，先不论译文的其他部分，就从句中谓语"推进"的选择上，develop显然并不符合英文的搭配，proceed才是合适的选择。

4. 与宾语搭配一致

（1）秦朝第一位皇帝秦始皇于公元前221年统一六国。

这里的"统一"没有简单地译成unify，而应根据与宾语"国"state的搭配，选择conquer。

Qin Shihuang, the first emperor of the Qin Dynasty, conquered the other states in 221 BC.

（2）很多成功的人来自贫穷的家庭，说明一个人不应该被基因或者家庭教育限制。

A. Many successful people are from poor families, which means that individuals should not be restrained by gene.

B. The fact that many successful people are from poor families means that one's development is not limited by genetic factors and upbringing.

A句中的restrain一般是指行为和情绪上的自制，而不是gene的限制。

5.连动式谓语的英译

连动式的谓语是指两个或两个以上的谓语动词共用一个主语的结构，这些动词间可以是不同的关系，包括并列、条件、方式、承接、因果，假设、目的等关系。翻译时要分析几个动词间的关系，将主要动词译成谓语，其他动词译成各种从属结构。请看以下例子。

（1）去年我和弟弟回乡去看朋友。

Last year, I went home with my younger brother to visit our friends.

“回乡”处理成为译句的谓语，“去看朋友”则是前者的目的。

（2）我第二天就离开了我那个朋友，并不知道以后还有没有机会再看见他。

I said good-bye to this friend, not knowing if I could ever see him again.

这句中的“并不知道”采用了现在分词，表示一种伴随的状态。

（3）他站起身，走到门口，把它扔了出去。

He rose up, walked to the door and threw it out.

原句中这3个动词是典型的并列关系。

Exercising Your Skills

Translate the following sentences into English using proper translation techniques.

1. 他到美国去攻读企业管理硕士了。

____

2. 缂丝可以在颜色之间进行顺畅的转换。

____

3. 我始终惦记着困难群众。

____

4. 无数宽敞美丽的庭院置身于屋舍之间。

____

5. 我的腿麻了。

____

6. 我的脑子里充满了忧患。

____

Unit Project

There are lots of silk brands in China, including some time-honoured Chinese brands like Ruifuxiang, Shengxifu, and Neiliansheng. Ruifuxiang specializes in making high quality silk. Predominantly making caps, Shengxifu, a ninety-year-old shop, is famous for its selection of material and handmade craftsmanship. And Neiliansheng, established in the Qing Dynasty, is famous for its Chinese cloth shoes.

Work with your partner on the following tasks:

1. Visit some silk shops in your city. Pick up one silk brand and collect information of different aspects of that brand, including its history, features, sales, target customers and so on. Report what you have found to the class.

2. Suppose your team will propose a long-term strategy of brand promotion for that silk company, write a plan and then show your plan in PPT before the class.

Unit 6 Keys

Unit 7 Cities and People

Unit Guide: Silk is more than a commodity. It's an envoy of Chinese culture. In **Passage 1** of this unit, we will meet face-to-face with a brocade master to understand the challenges that craftsmen face in promoting their ancient craft to the world, and in **Passage 2**, we will look at the efforts of Hangzhou, home of silk, to "re-lux" its image through silk, a key cultural product with huge economic value. As we strive to boost China's cultural influence abroad, internationally renowned fashion brands with distinctive Chinese flair might be the key to "Made in China 2025."

Lead-in

I Cities related to silk.

Step One: Fill in each blank with the name of the city that matches the picture and its tourist attraction.

Temple of Marquis Wu

1. ____________________

Canton Tower

2. ____________________

Humble Administrator's Garden

3. ____________________

Yuelu Academy

4. ____________________

West Lake

5. ____________________

Dr. Sun Yat-sen Mausoleum

6. ____________________

Step Two: Talk with your partners about what you know about these tourist attractions and in what way these cities are related to silk.

II Important people on the Silk Road.

Step One: In the history of the Silk Road, many renowned people left their footprints on this most historically important trade route. The following are some of those eminent people. Fill in the blanks with the correct names based on the profiles.

Xuanzang	Wang Zhaojun	Zhang Qian	Marco Polo	Ban Chao

1. __________ was a Chinese official and diplomat who served as an imperial envoy to the world outside of China in the late 2nd century BC during the reign of the Han Emperor Wudi of the Western Han Dynasty (206 BC—AD 8), opening up the ancient Silk Road and bringing reliable information about Central Asia and West Asia.

2. ________ was a famous general leader and diplomat in the Eastern Han Dynasty (25—220). Living in the Western Regions for about 31 years, he put down numerous rebellions and built diplomatic relations with more than 50 states, which guaranteed the long-lasting peace and harmony along the Silk Road.

3. __________ was a Chinese Buddhist monk, scholar, traveller, and translator who set out on his journey to India to study Indian sutras along the Silk Road in the 7th century and described the interaction between Chinese Buddhism and Indian Buddhism during the Tang Dynasty (618—907).

4. __________ was an Italian merchant, explorer, and writer, born in the Republic of Venice. In the Yuan Dynasty (1206—1368), he came to China and stayed for 17 years. His travels are recorded in *Book of the Marvels of the World*, a book that described to Europeans the wealth and great size of China, its capital Beijing, and other Asian cities and countries.

5. __________ was known as one of the Four Beauties of ancient China. Born in the Western Han Dynasty (206 BC—AD 8), she was sent by Emperor Yuan to marry the Xiongnu Chanyu (单于) Huhanye (呼韩邪) in order to establish friendly relations with the Han Dynasty through marriage. She was one of those princesses who made important contributions to the smooth flow of the Silk Road by the intermarriage.

Step Two: Time Travel.

If you have a chance to travel back to the past and be one of the above great people, which one would you like to be? And why?

Ⅲ Codes of the new Silk Road.

Watch the video and answer the following questions.

视 频

1. Why did Valencia become the largest city in Spain in the 15th century?

____

____

2. What was Valencia named by the United Nations Educational, Scientific and Cultural Organization in 2016?

____

____

3. What would the curator tell people from all over the world every day?

____

____

4. To smooth the silk trade exchanges, the exchange built a special court. What was the court used for?

____

____

5. What did Charlie's family depend on for generations? What does Charlie use it as now?

____

____

6. Why does the Valencia municipal government of Spain hope to seize the opportunities of "The Belt and Road Initiative" proposed by China?

____

____

Passage 1

Face-to-Face with a Silk Brocade Master

Silk has been a cultural treasure for China for around 5,000 years. Not only is it **revered** by millions of Chinese, but it also remains highly **sought after** by the western world, which has looked to import Chinese silk since the Han Dynasty (206 BC—AD 220) through the ancient land and **maritime silk routes.**

Brocade, which sees silk woven into **lavish** colourful patterns, **was reserved for** the nobles in ancient times. Its unique and extremely complicated weaving **process** made it a status symbol for those who could afford this luxury.

However, modern times have seen traditional Chinese brocading techniques almost disappear from the fashion world. The craft of brocading can no longer satisfy increasing consumer demands for **vibrant** colours and soft texture.

For the thousands of Chinese brocade craftsmen working today, getting their craft back on the shelves and passing on their traditional and highly-skilled work techniques remain an enormous challenge.

Li Jialin[1] is one of several master brocade craftsmen who have **spared no effort** to promote the craft, and **religiously** try to **steer** it towards a more prosperous future.

According to him, Chinese culture has always been rooted in the silk industry, which means that modern embroidery still borrows heavily from the traditional silk craftsmanship.

Li has amazed the world with his **intricate** craftsmanship on works including a woven silk version of the Song painting ***Along the River during the Qingming Festival***[2], as well as a brocaded copy of ***Sun Tzu's The Art of War***[3].

World's first silk copy of The Art of War ***debuted*** *in 2005 and was presented by former President Hu Jintao as a gift to former President* ***George Bush*** *in 2006.*

Li shared his views on the **renaissance** of silk embroidery in an **exclusive** interview with **CCTV NEWS.**

As a key cultural product, silk, which has long been popular overseas, can on one hand help improve China's cultural influence abroad, while on the other hand, provide an economic value as it competes with international fashion brands with its distinctive Chinese **flair** giving it a key advantage in the market.

Li spoke about how designs for brocade patterns on the qipao, a body-hugging traditional Chinese dress also known as a **cheongsam**, can draw inspiration from Chinese culture.

"We were once deeply impressed by the fascinating landscape of Hangzhou's Jiuxi (a series of nine streams and 18 gorges in the city), where the stream slowly flowed into the distance and a bunch of butterflies danced in the air chasing the spinning peach blossom petals, so we **contrived** a special qipao with patterns to recreate the scene," said Li, further **indicating** that traditional embroidery aims to represent a beautiful, romantic and typically Chinese scene.

While Chinese cultural elements should be applied to modern brocade, innovation in terms of design and technology to help push forward traditional embroidery is also of great importance to Li.

Li pointed out that Western fashion philosophy should be introduced into Chinese brocading techniques, so that the craft can **cater to** different fashion tastes. "Take the qipao as an example," he said, "besides the classical pattern of the '**dragon and phoenix**' [4] and '**blooming flower and full moon**[5], ' we could add some **chic** designs or western fashion elements as well to make it richer in style."

Innovation in design, however, can only be realized with modern technology.

Although Chinese brocading techniques were **unparalleled** during the late Ming and early Qing dynasties, where even gold threads and peacock feathers could be used to make fabrics, the traditional craft on its own is unable to meet the modern world's demands for a huge range of textiles and fashion goods.

Li Jialin has worked **strenuously** over the past 20 years to bring brocading into the 21st century, and his invention of the award-winning **Digital Emulative Colour Silk Weaving Technique** has proved to be a big step forward.

Employing computer technology in both the design and manufacturing process has enabled silk weaving to produce an even greater variety of patterns, from landscape paintings to works based on photography, using some 4,500 colours. This development has also seen a huge improvement in the **density** of brocading fabric, making it even softer.

*This **portrait** made of silk was gifted to the President of Poland **Andrezj Duda** and his wife when President Xi Jinping visited Poland in 2016.*

Li's use of modern technology has also allowed craftsmen to **customize** the entire process to suit their needs, all the way from weaving the silk to tailoring patterns based on individual demand.

"While our **customization** started from a piece of cloth in the past, it now starts from a thread," noted the master craftsman.

Through preserving Chinese cultural elements and constantly seeking new breakthroughs in both design and technology, "China should establish its own high-end brands for silk products to earn a position in the fashion world amid fierce competition with other international trademarks," said Li, talking about his expectations for the future of the country's silk industry.

"China has no lack of low and medium-grade goods as the world factory," added Li, "but what we should strive to do is to create our own '**Louis Vuitton**[6]' or 'Hermes[7].' "

With the 21st century **"The Belt and Road Initiative"** [8] now **in motion**, this precious material—whose **prevalence** along the ancient trade routes between east and west **conferred** the name the "Silk Road"—is now more than just a commodity.

Silk is an **envoy**, with the historic mission of promoting Chinese culture and bringing east and west closer. "While China's economy has grown quickly, its culture should also be boosted," Li said.

As one of the best representatives of the Chinese culture, "Chinese brocaded silk in modern times must reach out to the world."

Words

revere [rɪˈvɪə(r)] *vt.* 崇敬；尊崇

maritime [ˈmærɪtaɪm] *adj.* 海上的；海事的；航海的

lavish [ˈlævɪʃ] *adj.* 大量生产的；耗资巨大的

reserved [rɪˈzɜːvd] *adj.* 预订的；保留的

process [ˈprəʊses] *n.* 工序；工艺流程

vibrant [ˈvaɪbrənt] *adj.* 鲜艳的；醒目的；充满生机的

religiously [rɪˈlɪdʒəsli] *adv.* 笃信地；虔诚地

steer [stɪə(r)] *vt.* 引导；操纵；控制

intricate [ˈɪntrɪkət] *adj.* 错综复杂的；盘错的

debut [ˈdeɪbjuː] *vi.* 初次表演；初次登台

renaissance [rɪˈneɪsns] *n.* （某一学科或艺术形式等衰落后的）复兴

exclusive [ɪkˈskluːsɪv] *adj.* 专用的；排外的

flair [fleə(r)] *n.* 天资；鉴别力

cheongsam [tʃɒŋˈsæm] *n.* 旗袍

contrive [kənˈtraɪv] *vt.* 设计；创造；设法做到

indicate [ˈɪndɪkeɪt] *vt.* 表明；指示

cater [ˈkeɪtə(r)] *vt.* 满足需要；适合；投合，迎合

phoenix [ˈfiːnɪks] *n.* 凤凰

chic [ʃiːk] *adj.* 别致的；漂亮的；时髦的

unparalleled [ʌnˈpærəleld] *adj.* 空前的；无比的；无双的

strenuously [ˈstrenjʊəslɪ] *adv.* 奋发地；勤奋地

emulative [ˈemjʊlətɪv] *adj.* 仿真的；好胜的；竞争的

density [ˈdensəti] *n.* 密度

customize [ˈkʌstəmaɪz] *v.* 定制；定做

customization [ˈkʌstəmaɪzeɪʃən] *n.* 用户化；定制

portrait [ˈpɔːtreɪt] *n.* 肖像；肖像画

prevalence [ˈprevələns] *n.* 流行；盛行

confer [kənˈfɜː(r)] *vt.* 授予；颁与

envoy [ˈenvɔɪ] *n.* 使节；外交官

Useful Expressions

seek after 寻求；追求（过去式sought after）

maritime silk routes 海上丝绸之路

be reserved for 留作；（专）供……之用

spare no effort 不遗余力；尽力去做

cater to　迎合

blooming flower and full moon　花好月圆

in motion　在开动中；在运转中

Proper Names

Along the River during the Qingming Festival　《清明上河图》

Sun Tzu's The Art of War / The Art of War　《孙子兵法》

George Bush　乔治·赫伯特·沃克·布什（George Herbert Walker Bush，1924.6.12—2018.11.30），美国第51届第41任总统，常被称为老布什

CCTV NEWS　中国中央电视台英语新闻

Digital Emulative Colour Silk Weaving Technique　数码仿真色彩丝织技术

Andrezj Duda　安杰伊·杜达，波兰总统

Louis Vuitton　路易威登

The Belt and Road Initiative　"一带一路"倡议

Notes

1. Li Jialin is a professor and doctoral supervisor at the School of Art and Design of Zhejiang Sci-Tech University. He is the chief designer of Hangzhou Li Jialin Brocade Art Co., Ltd and Hangzhou Zhongfang Technology Development Co., Ltd, and enjoys special allowances from the State Council. His silk brocade paintings are collected in the National Museum of China (formerly the Museum of Chinese History), Beijing Palace Museum, National Library of China, China National Silk Museum, etc.
   李加林，浙江理工大学艺术与设计学院教授、博士生导师，杭州李加林织锦艺术有限公司、杭州中纺技术开发有限公司总设计师，享受国务院特殊津贴专家；他的丝绸织锦组画作品收藏于中国国家博物馆（原中国历史博物馆）、北京故宫博物院、中国国家图书馆、中国丝绸博物馆等。

2. *Along the River During the Qingming Festival* is a painting by the Northern Song Dynasty (960—1127) artist Zhang Zeduan (1085—1145). It captures the daily life of people and the landscape of the capital, Bianjing (present-day Kaifeng) during the Northern Song Dynasty. The theme is often said to celebrate the festive spirit and worldly commotion at the Qingming Festival, rather than the holiday's ceremonial aspects, such as tomb sweeping and prayers. The painting is considered to be the most renowned work among

all Chinese paintings.

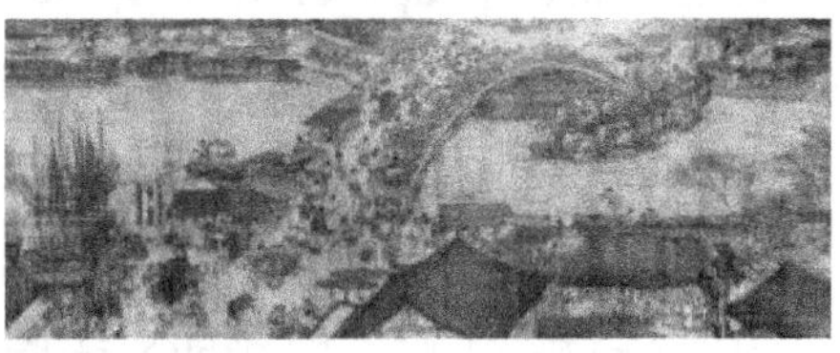

《清明上河图》，中国十大传世名画之一，北宋（960—1127）画家张择端（1085—1145）仅见的存世精品，属国宝级文物，现藏于北京故宫博物院。作品以长卷形式，生动记录了中国十二世纪北宋都城汴京（又称东京，今河南开封）的城市面貌和当时社会各阶层人民的生活状况，是北宋时期都城汴京当年繁荣的见证，也是北宋城市经济情况的写照。

3. *Sun Tzu's The Art of War* (also known as *The Art of War*) is an ancient Chinese military treatise dating from the Late Spring and Autumn Period (roughly 5th century BC). The work, which is attributed to the ancient Chinese military strategist Sun Tzu (Master Sun, also spelled Sunzi), is composed of 13 chapters, and each one is devoted to an aspect of warfare and how it applies to military strategy and tactics. It remains the most influential strategy text in East Asian warfare and has influenced both Eastern and Western military thinking, business tactics, legal strategy and beyond.

《孙子兵法》又称《孙武兵法》《吴孙子兵法》《孙子兵书》《孙武兵书》等，是中国现存最早的兵书，也是世界上最早的军事著作，被誉为“兵学圣典”。本书一共13篇，乃春秋时祖籍齐国乐安的吴国将军孙武所著。它在我国古代军事学术和战争实践中，都起过极其重要的指导作用。如今，《孙子兵法》被翻译成多种语言，在世界军事史上也具有重要的地位。

4. Deeply rooted in Chinese culture, dragon and phoenix were regarded as the most sacred animals and used to be emblems of emperor and empress. The Chinese dragon is traditionally the embodiment of the concept of yang (male), while phoenix is paired (yin, female) with dragon. Nowadays, dragon and phoenix often serve in classical Chinese art and literature as metaphors for people of high virtue and rare talent or, in certain combinations, for matrimonial harmony or happy marriage.
在中国传统观念中，龙和凤是最神圣的动物，曾经是皇帝和皇后的象征。中国龙传统上是“阳”（男性）概念的化身，而凤（“阴”，女性）与龙配对。如今，“龙凤”经常在中国古典艺术和文学中作为高尚美德和稀有人才的隐喻，或者代表着吉祥如意，多表示婚姻幸福。

5. The phrase “blooming flower and full moon” describes the scenery of the blooming flower and full moon, which is a metaphor for a good life in China, sometimes used to

congratulate people on their wedding or housewarming. The theme of "blooming flower and full moon" often appears in Chinese sculptures, paintings, folk music, movies, music, etc.
“花好月圆”，描述花开月圆的景色，比喻生活美好圆满，有时用于祝贺新婚或乔迁之喜。中国的雕刻品、绘画、民乐、电影、音乐等也常选用“花好月圆”为题材来创作。

6. Louis Vuitton, or shortened to LV, is a French fashion house and luxury retail company founded in 1854 by Louis Vuitton. The label's LV monogram appears on most of its products, ranging from luxury trunks and leather goods to ready-to-wear, shoes, watches, jewellery, accessories, sunglasses and books. It is one of the world's leading international fashion houses.
路易威登，或简称为LV，是一家法国时尚家居和奢侈品零售公司。自1854年以来，代代相传至今的路易威登，以卓越品质、杰出创意和精湛工艺成为时尚旅行艺术的象征。产品包括手提包、旅行用品、小型皮具、配饰、鞋履、成衣、腕表、高级珠宝及个性化订制服务等。

7. Hermès International S.A., or simply Hermès, is a French high fashion luxury goods manufacturer established in 1837. It specializes in leather, lifestyle accessories, home furnishings, perfumery, jewellery, watches and ready-to-wear. Its logo, since the 1950s, is of a Duc carriage with horse.
爱马仕是世界著名的奢侈品品牌，1837年由蒂埃里·爱马仕创立于法国巴黎，早年以制造高级马具起家，迄今已有180多年的悠久历史。爱马仕一直以精美的手工和贵族式的设计风格立足于经典服饰品牌的巅峰。

8. The Belt and Road Initiative (BRI), also known as the Silk Road Economic Belt and the 21st-century Maritime Silk Road, is a development strategy adopted by the Chinese government involving infrastructure development and investments in countries in Europe, Asia and Africa. "Belt" refers to the overland routes for road and rail transportation, called "the Silk Road Economic Belt" ; whereas "road" refers to the sea routes, or the 21st Century Maritime Silk Road.
“一带一路”是“丝绸之路经济带”和“21世纪海上丝绸之路”的简称，2013年9月和10月由中国国家主席习近平分别提出建设“新丝绸之路经济带”和“21世纪海上丝绸之路”的合作倡议。“一带一路”倡议旨在借用古代丝绸之路的历史符号，积极发展与沿线国家的经济合作伙伴关系，共同打造政治互信、经济融合、文化包容的利益共同体、命运共同体和责任共同体。

Reading Comprehension

Complete the sentences with the information given in the passage.

1. Since the Han Dynasty, Chinese silk has exported to the Western world through the Silk Road and ____________________.

2. In ancient China, brocade, with silk woven into lavish colourful patterns, was a ____________________ for the nobles.

3. Modern times have witnessed ____________________ for bright colours and soft texture which the craft of traditional Chinese brocading can no longer satisfy.

4. Li Jialin's brocaded copy of *Sun Tzu's The Art of War* has impressed and astonished people with his____________________.

5. According to Li, while modern brocade should preserve Chinese cultural elements, it's also important for traditional embroidery to be pushed forward by innovation in ____________________.

6. Li considers that Chinese brocading techniques can be combined with ____________________ so as to meet different needs and tastes of people around the world.

7. Li Jialin's invention of the award-winning computer technology ____________________ has made innovation in design possible.

8. In a fiercely competitive international market, China should endeavour to create its own high-end brands for silk products rather than only ____________________ as the world factory.

9. ____________________ derived its name from the phenomenon that silk was very common and popular along the ancient trade routes between east and west.

10. Silk is now rather an envoy than just ____________________, with the historic mission of promoting Chinese culture and bringing east and west closer.

Summary

The passage points out the problems that traditional Chinese brocading faces in modern times, namely nearly disappearance of its techniques from the fashion world and its craft's inability to meet the modern world's demands for vibrant colours and soft texture. With regard to it, the passage shares Li Jialin's views on the renaissance of silk embroidery. Fill in the blanks with appropriate words.

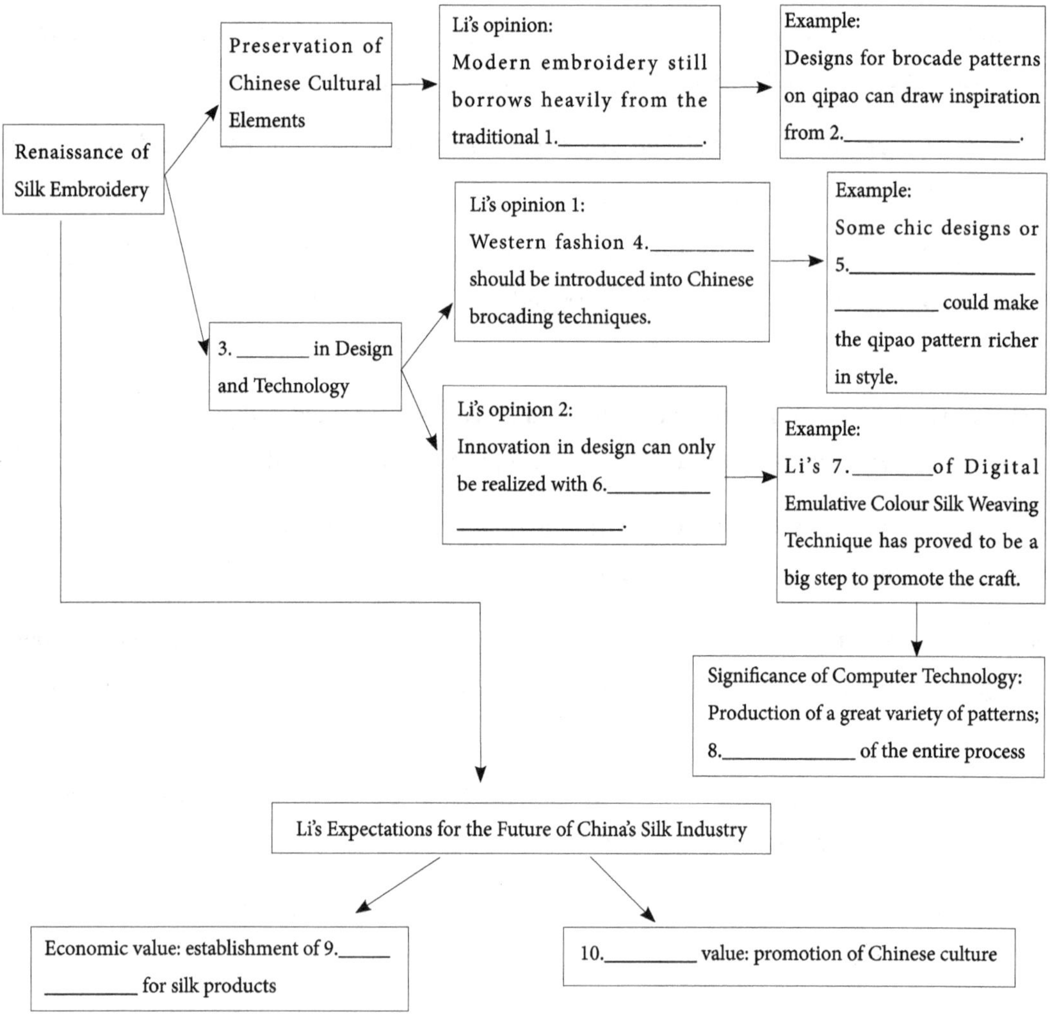

Renaissance of Silk Embroidery

→ Preservation of Chinese Cultural Elements

→ Li's opinion: Modern embroidery still borrows heavily from the traditional 1.____________.

→ Example: Designs for brocade patterns on qipao can draw inspiration from 2.____________.

→ 3._______ in Design and Technology

→ Li's opinion 1: Western fashion 4._________ should be introduced into Chinese brocading techniques.

→ Example: Some chic designs or 5.____________ _________ could make the qipao pattern richer in style.

→ Li's opinion 2: Innovation in design can only be realized with 6._________ ____________.

→ Example: Li's 7._______of Digital Emulative Colour Silk Weaving Technique has proved to be a big step to promote the craft.

→ Significance of Computer Technology: Production of a great variety of patterns; 8.____________ of the entire process

→ Li's Expectations for the Future of China's Silk Industry

→ Economic value: establishment of 9._____ _________ for silk products

→ 10._________ value: promotion of Chinese culture

Word Building

I The prefix *circum* can be used together with other words to form new words, meaning "around, about." Match the following words with their parts of speech and Chinese meanings.

	circumaviation	环流
	circumference	规避；绕行
	circumnavigate	迂回曲折的陈述
n.	circumscribe	状况；条件；境遇
	circum-continental	极圈的；极地附近的
adj.	circumvent	圆周；圆周长
	circumlocution	环球航行
v.	circumstances	限制
	circumfluence	环球飞行
	circumpolar	环大陆的

II The root word *mari* can be used together with prefix and / or suffix to form new words, meaning "sea." Match the following words with their parts of speech and Chinese meanings.

	marine	海运的，海事的
	mariner	横渡海洋的
	aquamarine	海上飞行的
n.	submarine	海蓝宝石；海蓝色
	maritime	海的
adj.	mariculture	海潮记录仪；验潮计
	marinate	用盐水泡；腌
v.	aeromarine	海底的；潜水艇
	transmarine	船员；水手
	marigraph	海水养殖

Words and Expressions in Use

I Fill in the blanks with the words given below. Change the form where necessary. Each word can be used only once.

strenuously	lavish	confer	contrive	customize	revere
prevalence	vibrant	cater	envoy	unparalleled	flair

1. We have to __________ a very large black-out curtain to ensure the success of the experiment.

2. With the fall in full swing, Shanghai Disney Resort invites guests from around the world to immerse themselves in the resort's _________ autumn colours and seasonal experiences.

3. Thanks to the __________ of smart devices and digital payment systems, consumers are already getting used to ordering products and services through online channels. They also expect prompt delivery of quality goods and services.

4. __________ scenes and costumes combine to depict a girl's fantasy adventure in a magical kingdom after she meets a toy nutcracker in tonight's ballet show "The Nutcracker" based on E.T.A Hoffmann's story.

5. In the history of Chinese modern painting, Huang Binhong enjoyed equal popularity with Qi Baishi. They two were dubbed "Northern Qi and Southern Huang" in history, a reflection of their __________ status in modern art.

6. Peking Opera has been __________ as the opera of China and an icon of traditional Chinese culture.

7. Henry Fok Ying-tung devoted __________ efforts to helping the nation gain membership of a number of international sporting organizations that were important to the rise of modern China.

8. Peng Liyuan, wife of Chinese President Xi Jinping and a UNESCO Special __________ for the Advancement of Girls' and Women's Education, visited a pre-school in the South African capital of Pretoria on Tuesday, July 24, 2018.

9. The newly-opened restaurant is celebrating the season with an Italian __________, blending local ingredients and seasonal imported produce from Italy.

10. China opened its first Internet court in Hangzhou, capital of East China's Zhejiang Province, last August to __________ to the increasing number of online disputes.

11. People in China have demonstrated their zeal to buy foreign goods that they deem to be of better quality or that they think __________ more social status.

12. From 2016 to 2017, the trend of decorating lipsticks with the Palace Museum Taobao's beautiful tapes grew vastly popular in China. Numerous women started to __________ their makeup collection creatively to embrace Chinese culture and heritage.

II Render the following Chinese expressions into English. The first letter of each word has already been given.

1. 旋转的桃花瓣 s________ p________ b________ p________
2. 独特的中国风情 d________ C________ f________
3. 重新上架 b________ o________ t________ s________
4. 走向世界 r________ o________ to t________ w________
5. 独家专访 e________ i________
6. 高端品牌 h________ b________
7. 大师级锦缎工匠 m________ b________ c________
8. 花好月圆 b________ f________ and f________ m________
9. 中低档商品 l________ a________ m________ g________
10. 定制模式 t________ p________

Translation

Translate the following paragraph into English.

苏绣成绩斐然，拥有一批技艺精湛的艺术家。20世纪初期，在中国东部有一位漂亮的女性，她因精美的刺绣享誉全国，其后在全世界赫赫有名。她就是被清朝著名学者俞樾形容为“针神”的苏绣大师沈寿（1874—1921）。现在，她精美的绣品在北京、南京、上海、苏州和南通以及其他国家的博物馆展出。当人们看到她的刺绣作品时，无不为之倾倒，惊叹她精湛的技艺。她凭借非凡的智慧和灵巧的绣手，将苏绣艺术提升到了一个高度，创造了许多无与伦比的杰作。

Passage 2

Made in China: The Silk Road to Soft Power

Amber Butchart[1]

How Hangzhou—home to the China National Silk Museum and ***Alibaba—is*** *becoming* ***synonymous with*** *China's desire to re-lux*[2] *its image.*

"China Through the Looking Glass", 2015, ***installation view, Blue and White Porcelain*** *room; Photograph: ©* ***The Metropolitan Museum of Art****, New York*

Hangzhou in Zhejiang Province in East China has been a centre of sericulture for centuries and is home to the China National Silk Museum. One of the largest museums of dress and textiles in the world, it is also home to **the Key Scientific Research Base of Textile Conservation** as well as a number of **initiatives** that focus on the study of Silk Road **material culture**. Hangzhou itself was the capital city during part of the Song Dynasty (960—1279) and has long been considered a site of natural beauty and **contemplation**. **Steeped** in **folklore**, the city is the setting for fables such as "the Legend of the White Snake[3]," one of China's four great folktales. Its landscape is dominated by the **tranquillity** of the West Lake and its islands, surrounded by **undulating** hills and dotted with temples and **pagodas**. Since the 9th century CE it has been celebrated by poets and artists, and the lake's features were imitated for Kunming Lake at Beijing's Summer Palace, the **retreat** for the last imperial family, the Qing Dynasty. An old Chinese proverb celebrates the wonders of the scenery, "In Heaven there is Paradise, on earth Suzhou and Hangzhou." [4]

Silk is woven into Hangzhou's economic and artistic history, from the **Hangzhou Weaving Bureau** which ran the imperial silk workshops during the Ming Dynasty (1368—1644), to

The First Monday in May[5], *2015, film still. Courtesy: Magnolia Pictures*

the "**Hall of Sericulture Studies**," the first institution dedicated to modern sericulture education in 1897. In the 1920s, factories in Zhejiang Province were at the forefront of the **mechanization** of silk production as the process moved from handcraft to industry. Yet Hangzhou's fortunes are not all based in the past. It is also the home to Alibaba, China's e-commerce and finance **behemoth** which established the city as a tech centre that has been **likened to Silicon Valley** in the US. Brands based in the town are becoming aware of the economic power of **monetizing** heritage. Hangzhou-based silk company **Wensli** have been operating for more than 40 years, and chairwoman **Tu Hongyan** claims a family connection in Hangzhou's silk industry dating back to the Southern Song Dynasty (1127—1279). Wensli silk scarves are a favourite of politicians and **statespeople** looking to include a touch of diplomacy in their dress. **Christine Lagarde, Managing Director of the International Monetary Fund**, wore a Wensli scarf during the **G20 Hangzhou Summit** in 2016. Looking to compete in an international market, in 2013 the Wensli Group **acquired Marc Rozier**, a French silk scarf factory with a 120-year history of manufacturing for brands such as Hermes. This move **illuminates** broader histories concerning shifts in taste and production: if silk's origins are in China and cities like Hangzhou, Paris remains the home of **couture** and the symbolic production of luxury.

It is the concept of luxury that Chinese government is looking to readdress. The Silk Road, a series of interconnecting routes of trade and cultural exchange that joined the Far East to the **Mediterranean**, was the key to China's fortune throughout **antiquity** until the **fragmentation** of the **Mongol Empire** in the 14th century. Chinese products were synonymous with luxury—even reaching Roman emperors—and the creation of products such as silk and porcelain remained closely guarded secrets. Conceptually, the fortunes of the Silk Road remain a **bedrock** of Chinese policy—the Great Wall of China was extended to protect it. Launched in 2013, the Belt and Road Initiative aims to revive those ancient **freight** networks and establish new ones between Asia, Europe and East Africa, and while critics remain divided on the impact of the network on the global fashion industry, the Yiwu-London train route[6] has already opened. Likewise, the "Made in China 2025"[7] programme is a government-backed project to shift the economy towards advanced manufacturing sectors including robotics, ocean engineering, satellites and high-tech transportation, among others, speaking to the desire to **reclaim** the "Made in China" tag and have it stand for quality and innovation rather than **budget mass production**. It has **ruffled** enough **feathers** to have been criticized by **Donald Trump** and cited in his threats of a trade war between the U.S. and China.

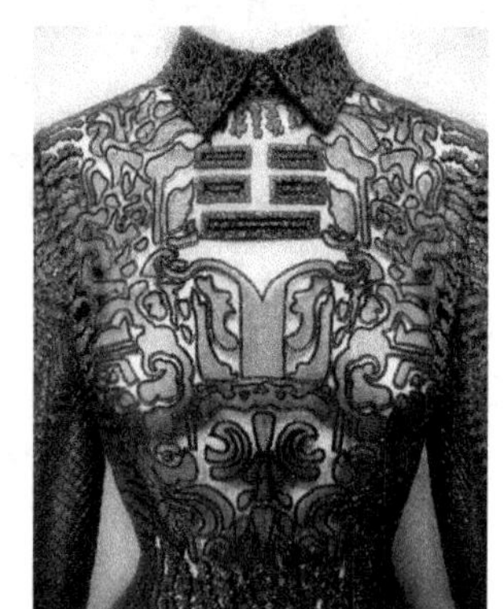

Valentino SpA**, Evening dress, 2013, "Shanghai" collection. Courtesy: © Valentino SpA; photograph: **Platon Antoniou

Historically, viewing fashion as a form of soft power can help to **unpick** the cross-cultural impact of Chinese **aesthetics** on European design and **vice versa**, as well as the shifting cultural influence of nations. China is the world's second largest economy (by **nominal GDP**), and remains a leading producer and exporter of silk. Recognition can be found in **blockbuster** museum shows such as "China: Through the Looking Glass," a collaboration between the **Costume Institute** and the **Department of Asian Art** at New York's Metropolitan Museum in 2015. The exhibition was the Met's fifth most-visited show ever, topping 2011's "**Alexander McQueen: Savage Beauty**" which was the Costume Institute's previous bestseller. "China: Through the Looking Glass" focused on China as a source of inspiration to the West and had to negotiate a fine line between celebrating and **perpetuating reductive** notions of Orientalism, a thread which was woven through the **documentary** by **Andrew Rossi** about the show's **curation**, *The First Monday in May* (2016). A previous show at **the V&A** in 2011, "**Imperial Chinese Robes from the Forbidden City**", saw many Qing Dynasty garments on display in Europe for the first time. Within China, an interest in rebuilding museum collections developed. The 1990s saw an acceleration of attempts to rebuild and reprioritize heritage (Hangzhou's China National Silk Museum opened in 1992), and museum collaborations now form part of the Belt and Road Initiative.

Concurrently, a new generation of designers and brands such as **Awaylee, C.J. Yao, Ms Min** and **Alexander T. Zhao** are **repositioning** attitudes towards Chinese design and luxury on both a domestic and international stage. **Established couturier Guo Pei** is known for celebrating the handwork element of her craft, regularly **clocking up** thousands of hours to create a single dress (including the one **Rihanna** wore to the **Met Gala**). Taking inspiration from Chinese material culture such as blue and white porcelain, she is redefining luxury using concepts from both China and Europe and in 2016 was invited to become a guest member of the **Chambre Syndicale de la Haute Couture** in Paris, the governing body of

French couture. As with Hangzhou, fashion in China points towards a **fusion** of celebrating the past and a future-focused **embrace** of the present. A **reappraisal** of the "Made in China" label might just be the beginning.

Words

synonymous [sɪˈnɒnɪməs] *adj.* 同义的；同义词的

porcelain [ˈpɔːsəlɪn] *n.* 瓷；瓷器

initiative [ɪˈnɪʃətɪv] *n.* 倡议

contemplation [ˌkɒntəmˈpleɪʃn] *n.* 沉思；冥想

steeped [stiːpt] *adj.* 充满……的；沉浸在……中的

folklore [ˈfəʊkˌlɔː(r)] *n.* 民间传说；民俗学

tranquillity [træŋˈkwɪləti] *n.* 安宁；平静

undulate [ˈʌndjʊleɪt] *vi.* （使）起伏；（使）波动；（使）呈波浪形

pagoda [pəˈɡəʊdə] *n.* 塔；宝塔

retreat [rɪˈtriːt] *n.* 僻静处；隐居处

bureau [ˈbjʊərəʊ] *n.* 办事处；机构

mechanization [ˌmekənaɪˈzeɪʃn] *n.* 机械化

courtesy [ˈkɜː(r)təsi] *n.* 承蒙；谦恭有礼，礼貌

magnolia [mæɡˈnəʊliə] *n.* 木兰花；木兰

behemoth [bɪˈhiːmɒθ] *n.* 超级公司（或机构）

liken [ˈlaɪkən] *v.* 比作

silicon [ˈsɪlɪkən] *n.* 硅

monetize [ˈmʌnɪtaɪz] *vt.* 把……作为法定货币；使具有货币性质；把……铸成货币

statesperson [ˈsteɪtsˌpɜː(r)s(ə)n] *n.* 政治家 [pl.] statespeople

acquire [əˈkwaɪə(r)] *vt.* 购得，得到

illuminate [ɪˈluːmɪneɪt] *vt.* 照亮；阐明

couture [kuˈtjʊə(r)] *n.* 时装设计制作；时装

Mediterranean [ˌmedɪtəˈreɪniən] *n.* 地中海
adj. 地中海的

antiquity [ænˈtɪkwəti] *n.* 古老；古代

fragmentation [ˌfræɡmenˈteɪʃn] *n.* 分裂；破碎；碎裂

bedrock [ˈbedrɒk] *n.* 基石；基础；根本原则

freight [freɪt] *n.* 货运；货物；运费

reclaim [rɪˈkleɪm] *vt.* 取回；要求归还

budget [ˈbʌdʒɪt] *adj.* 价格低廉的；花钱少的

ruffle [ˈrʌfl] *vt.* 激怒；弄皱

metropolitan [ˌmetrəˈpɒlɪt(ə)n] *adj.* 大城市的；大都会的

unpick [ˌʌnˈpɪk] *vt.* 深入剖析（复杂问题或疑惑）

aesthetics [iːsˈθetɪks] *n.* 美学；审美学

nominal [ˈnɒmɪn(ə)l] *adj.* 名义上的

GDP [ˌdʒiːdiːˈpiː] *abbr* 国内生产总值（全写为 gross domestic product）

blockbuster [ˈblɒkbʌstə(r)] *n.* 一鸣惊人的事物；（尤指）非常成功的书（或电影）

savage [ˈsævɪdʒ] *adj.* 野蛮的

perpetuate [pəˈpetʃueɪt] *vt.* 使保持；使持久化

reductive [rɪˈdʌktɪv] *adj.* 简化论的；简化的；以简释繁的

documentary [ˌdɒkjʊˈment(ə)ri] *n.* 纪录片

curation [ˌkjuːˈreɪʃ(ə)n] *n.* 策展（即策划、筛选并展示的意思）

concurrently [kənˈkʌrəntlɪ] *adv.* 同时地

reposition [ˌriːpəˈzɪʃən] *vt.* 换位；改变位置（主张、态度、立场等）

established [ɪˈstæblɪʃt] *adj.* 著名的；成名的

couturier [kuˈtjʊərieɪ] *n.* 女服设计师；女时尚裁缝

fusion [ˈfjuːʒ(ə)n] *n.* 融合

embrace [ɪmˈbreɪs] *n.* 拥抱；接受

reappraisal [ˌriːəˈpreɪzl] *n.* 重新评价

Useful Expressions

be synonymous with 和……同义；……的意义和……是一样的

installation view 装置景观

blue and white porcelain 青花瓷器

(by) courtesy of 蒙……提供、赞助、赠送；承蒙……的允许（或好意）

material culture 物质文化

film still 剧照

liken sb. / sth. to sb. / sth. 把……比作……

mass production 大规模生产；批量生产

ruffle (someone's) feathers 激怒（某人）

vice versa 反之亦然

nominal GDP　名义GDP（也称货币GDP，是用生产物品和劳务的当年价格计算的全部最终产品的市场价值）

clock up　达到（总量，总数）

Proper Names

Alibaba / Alibaba Group Holding Limited　阿里巴巴集团控股有限公司

China: Through the Looking Glass　中国：镜花水月

The Metropolitan Museum of Art　大都会艺术博物馆（简称The Met）

The Key Scientific Research Base of Textile Conservation　纺织品文物保护重点科研基地

Hangzhou Weaving Bureau　杭州织造局

Hall of Sericulture Studies　养蚕研究中心

The First Monday in May　（时尚纪录片）《五月首周一》

Magnolia Pictures　美国木兰影业 / 美国木兰花影业

Silicon Valley　硅谷，位于美国加利福尼亚州，是计算机和电子公司聚集地

Wensli / Wensli Group　万事利集团

Tu Hongyan　屠红燕，浙江杭州人，现任万事利集团党委书记及董事长

Christine Lagarde　克里斯蒂娜·拉加德，法国前财政部部长及国际货币基金组织首位女性总裁

Managing Director of the International Monetary Fund　国际货币基金组织总裁

G20 Hangzhou Summit　G20杭州峰会

Marc Rozier　马可·罗尼，法兰西丝绸世家

Mongol Empire　蒙古帝国

Donald Trump　唐纳德·特朗普，1946年生于纽约，美国共和党籍政治家、第45任美国总统

Costume Institute　时装学院

Department of Asian Art　亚洲艺术部

Alexander McQueen　亚历山大·麦昆，英国著名设计师

Alexander McQueen: Savage Beauty　“亚历山大·麦昆：野性之美”，2011年美国大都会艺术博物馆时尚大展

Valentino SpA　华伦天奴，著名意大利高级时装品牌

Platon Antoniou　普拉顿·安东尼乌，英国著名摄影师

Andrew Rossi　安德鲁·罗斯，美国导演、编剧

The V & A　维多利亚和阿尔伯特博物馆，全称The Victoria and Albert Museum，英国乃至世界最大博物馆之一

Imperial Chinese Robes from the Forbidden City “来自紫禁城的中国皇室长袍”，由中国故宫博物院和伦敦维多利亚和阿尔伯特博物馆联合举办的服饰展（2010年12月7日—2011年2月27日）

Awaylee 李薇，著名的时装设计师，创立女装品牌Awaylee

C. J. Yao 著名时装设计师，出生在上海

Ms Min 刘旻，著名时装设计师，福州人，创立女装品牌MS Min

Alexander T. Zhao 由著名时装设计师赵涛（Alessandra Zhao）创立的独立品牌

Guo Pei 郭培，著名时装设计师，中式服装高级定制设计师

Robyn Rihanna Fenty 罗比恩·蕾哈娜·芬缇，在美国发展的巴巴多斯籍女歌手、演员、模特

Met Gala / Met Ball 纽约大都会艺术博物馆慈善舞会

Chambre Syndicale de la Haute Couture 法国高级时装协会

Notes

1. Amber Butchart is a fashion historian, author and broadcaster, who specializes in the historical intersections between dress, politics and culture. She was the presenter of BBC4’s six-part series *A Stitch in Time* that fused biography and art to explore the lives of historical figures through the clothes they wore. Her new publication, *The Fashion Chronicles: Style Stories of History's Best Dressed*, has been out in September, 2018.
   安伯·布查特是一位时尚历史学家、作家和广播员，专门研究服装、政治和文化之间的历史交集。她曾是BBC4频道六集英国时装简史记录片《历史的针脚》的主持人，该系列融合了传记和艺术，通过历史人物穿的衣服探索他们的生活。她的新书《时尚编年史：历史上最佳着装的风格故事》已于2018年9月份出版。

2. “re-lux”—definition of lux in *Merriam-Webster Dictionary*: a unit of illumination equal to the direct illumination on a surface that is everywhere one meter from a uniform point source of one candle intensity or equal to one lumen per square meter. Here in the passage, “re-lux” could mean “re-illuminate.”
   勒克司度（lux），即光照度，表示被摄主体表面单位面积上受到的光通量。此处re-lux可指“重新照亮”，用作动词，与image搭配，表示“重塑形象”。

3. “The Legend of the White Snake,” also known as “Madame White Snake,” is a Chinese legend, which is now counted as one of China’s Four Great Folktales, the others being “Lady Meng Jiang” “The Cowherd and the Weaving Maid” “ Liang Shanbo and Zhu Yingtai” (also known as “The Butterfly Lovers”).

《白蛇传》传说源远流长，是中国四大民间爱情传说之一（其余三个为《孟姜女》《牛郎织女》《梁山伯与祝英台》），描述的是一个修炼成人形的蛇精与人的曲折爱情故事。

4. "In Heaven there is Paradise, on earth Suzhou and Hangzhou" is a popular Chinese saying, praising two cities of Suzhou, Jiangsu Province, and Hangzhou, Zhejiang Province.
   "上有天堂，下有苏杭"是我国民间流传的谚语，意在赞叹江南美景，可与天堂相媲美。

5. *The First Monday in May* is a 2016 documentary film directed by Andrew Rossi. The film follows the creation of the Metropolitan Museum of Art's most attended fashion exhibit in history: the 2015 art exhibition "China: Through the Looking Glass" by curator Andrew Bolton at New York's Metropolitan Museum of Art, which, colloquially "The Met," is the largest art museum in the United States.
   《五月首周一》是安德鲁·罗斯导演的纪录片电影，该片于2016年4月15日在美国上映。该片讲述了2015年由策展人安德鲁·博尔顿（Andrew Bolton）策划的纽约大都会艺术博物馆"中国：镜花水月"为主题的时尚展览及博物馆慈善舞会的筹备与盛况。大都会艺术博物馆是美国最大的艺术博物馆，也是世界著名博物馆。

6. The Yiwu-London train route or the Yiwu–London Railway Line is a freight railway route from Yiwu, China, to London, United Kingdom, covering a distance of roughly 12,000 km (7,500 miles). This makes it the second longest railway freight route in the world after the Yiwu-Madrid railway line, which spans 12,874 km (8,000 miles). It is part of establishing a modern-day Silk Road. The route was opened on 1 January 2017, making London the 15th European city to have a railway route connection with China, and takes 18 days to complete.
   义乌至伦敦中欧班列是货运铁路线，从浙江义乌出发，途经亚洲和欧洲多国，到达终点伦敦。作为中英贸易直通车，该趟班列于2017年1月1日从中国义乌西站出发，全程运行约12000千米（7500英里），历时18天到达终点站，开启了中英贸易全新的物流通道，仅次于义乌至马德里铁路线，长达12874千米（8000英里），成为第二长铁路线。对于强化中国与西欧陆路互联互通，密切中英贸易往来，更好地服务"一带一路"建设具有重要意义。

7. "Made in China 2025" is a strategic plan of China issued by Chinese Premier Li Keqiang and his cabinet in May 2015. China is moving away from being the World's factory floor (cheap goods and low quality) to higher value products and services. In essence, it's a

blueprint to upgrade the manufacturing capabilities of Chinese industries.

“中国制造2025”是中国国务院总理李克强于2015年5月发布的中国战略规划。中国正在从世界工厂（廉价商品和低质量）转向高价值的产品和服务。从本质上讲，这是提升中国工业制造能力的蓝图。

Reading Comprehension

I Decide whether the following statements are true or false. Write “T” for true and “F” for false.

1. (　　) China National Silk Museum is located in Hangzhou, Zhejiang Province, which is the largest museum of dress and textiles in the world.

2. (　　) West Lake, whose features imitated Kunming Lake at Beijing’s Summer Palace, is one of Hangzhou’s tourist attractions.

3. (　　) Hangzhou is compared to be Silicon Valley, a tech centre in the U.S., because it is home to Alibaba Group.

4. (　　) The silk scarf Christine Lagarde wore during the G20 Hangzhou Summit was produced by Wensli, a Hangzhou-based silk company.

5. (　　) On the ancient Silk Road, Chinese products like silk and porcelain were considered to be luxury and how they were created was kept secret.

6. (　　) Yiwu-London train route connects China with the U.K., covering several countries in Asia and Europe, which is welcomed by the people along the route.

7. (　　) At present, “Made in China” tag stands for quality and innovation rather than cheap goods and low quality.

8. (　　) *The First Monday in May* was a documentary about the curation of Met’s show “Imperial Chinese Robes from the Forbidden City” in 2011.

9. (　　) As part of the Belt and Road Initiative, museums collaborate with each other to have some shows or exhibitions in China or western countries.

10. (　　) More and more Chinese fashion designers take Chinese material culture as source of inspiration to create their works with Chinese flair, trying to redefine “Made in China” label on both a domestic and international stage.

II Outline "Hangzhou as a centre of sericulture for centuries" in timeline pattern.

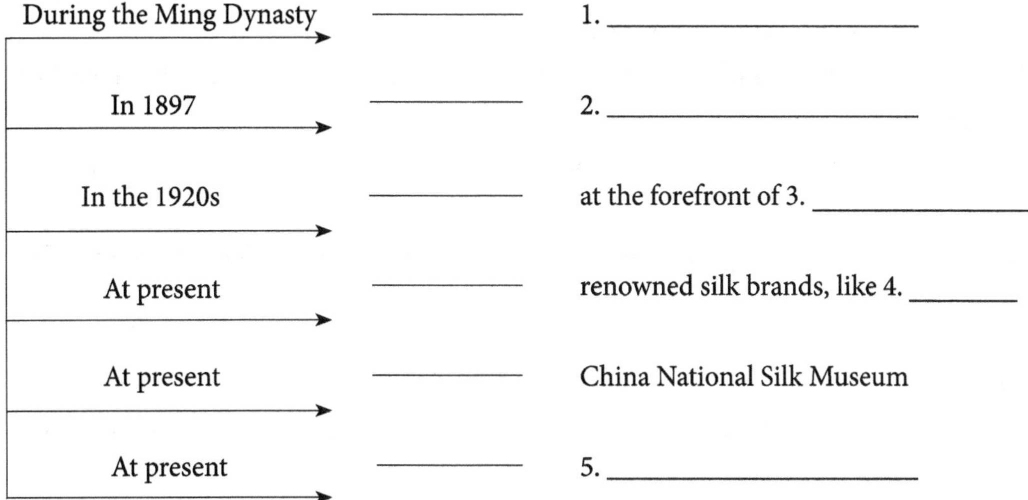

•• Words and Expressions in Use ••

I Match the collocations in Column A with the Chinese meanings in Column B.

Column A	Column B
1. mass production	A. 重塑形象
2. the Forbidden City	B. 政府支持的项目
3. guest member	C. 剧照
4. government-backed project	D. 起伏的山丘
5. re-lux one's image	E. 大规模生产
6. film still	F. 特邀代表
7. soft power	G. 紫禁城
8. undulating hills	H. 软实力

II Compare the following pairs of sentences and explain the different parts of speech and meanings of italicized words.

1. budget

(A) In families with children, it is usually the mother who manages the household ***budget***.

(B) The growth in ***budget*** airlines and cheap car hire means travel is not difficult.

(C) This is the first time we have had a series of ***budgets*** that has invested heavily in the economy, our social infrastructure, and our people.

(D) Another major issue is how to ***budget*** the money to pay for all the security needs.

2. illuminate

(A) Cross-cultural research can help ***illuminate*** and clarify the social and developmental differences that exist across various cultures.

(B) A great flash of lightning ***illuminated*** the world outside, showing the trees dark against the night sky.

(C) The museum purchased an important collection of medieval and renaissance ***illuminated*** manuscripts.

(D) During the Lantern Festival, most temples are ***illuminated*** by colourful lanterns of all shapes and sizes.

3. retreat

(A) More practically, it gives them a place to ***retreat*** to, escaping the stress of life in the public eye.

(B) In the summer of 1915 the Russian army ***retreated*** on its southwest front.

(C) Obviously the council has now ***retreated*** in the face of their united voice.

(D) He wanted a quiet ***retreat*** to build his house upon and concentrate on his work.

4. embrace

(A) Much has changed in photography over the last 50 years and the club has readily ***embraced*** these changes.

(B) Night grew darker with increased time, and the city stood peaceful at the ***embrace*** of the lights that surrounded it.

(C) The festival ***embraces*** various strands of the arts including music, theatre, dance, film, comedy, literature and family entertainment.

(D) He immediately stood up and rushed towards his friend, ***embracing*** him warmly.

5. reclaim

(A) This land was ***reclaimed*** from the sea about 2,000 years ago.

(B) This paper analyzes the ***reclaim*** of waste battery.

(C) He was determined to ***reclaim*** the British crown for his father.

(D) He is working to ***reclaim*** Maori prisoners by linking them to their racial traditions.

6. freight

(A) In the first five months of this year, 3.6 million tons of crude oil had been ***freighted*** to this area.

(B) Passengers stopped travelling that line in 1970 and ***freight*** trains stopped using it in 1980.

(C) Generally, our ornamental products are shipped unassembled to save ***freight***.

(D) Some shipments may be transferred from air to ocean ***freight*** as customers accept longer journey time to save money

Ⅲ Fill in the blanks with the words or expressions given below. Change the form where necessary. Each word or expression can be used only once.

blockbuster	established	monetize	courtesy of	synonymous	concurrently
clock up	reposition	vice versa	ruffle	bedrock	perpetuate

1. All the photos are ____________ our gracious holiday hosts, an added bonus to their amazing hospitality.

2. With more investment in the industry, more ____________ movies with stunning visuals have emerged.

3. For China-Africa economic contact to be sustainable, there should be ____________ an emphasis on cultural exchanges in the hope of attaining better understanding between the people.

4. This restaurant has been ____________ with the city's top steakhouses and gained its loyal diners with its prime first-grade quality steaks.

5. Some of those monuments are almost a century old and were erected to ____________ a memory and a spirit dear to surviving family members.

6. After ____________ 10,000 miles, the car still sounds as smooth as the day I first drove it.

7. They think this high profile meeting in London will ____________ his feathers.

8. In that backwater area, the easiest way to prosper is to ____________ the mountains or forests as mining or tourist resources.

9. You can only compose if you understand how to conduct and ______________.

10. Instant noodles were once the ______________ of China's convenience food, but their sales had declined drastically in recent years.

11. The clothing retailer has ___________ its brand to include more fashionable products.

12. Lin Ching-hsuan is considered to be one of the most prolific and ___________ contemporary essayists in China. His first book, *Lian Hua Kai Luo* or *Blossoming and Falling of Lotus* was published in 1973.

Translation Skills

句子英译（3）——句子其他成分的调整

前面章节提到英语句子主谓结构突出，有固定的基本句型，而且可以在主干基础上，借助连接词和语法手段，附加各种短语结构和从句结构，形成包含各种层次结构和逻辑关系的复杂长句。汉译英时，我们在确定了主谓之后，就要考虑句子其他成分的调整了，以符合英语的规范。

1. 状语位置的调整

汉语中的状语一般放置在主语之后，谓语之前，也有为了强调，放在主语之前的。英语中状语的位置则相对比较灵活，单词状语可以位于句首、句中、句末，较长的状语则常被置于句首或句末。因此汉译英时，壮语位置的调整也显得比较复杂。例如：

（1）在机器中很多能量是由于部件之间的摩擦而损失的。

In a machine a great deal of energy is lost because of the friction of its parts.

原文中有两个状语，“在机器中”和“由于部件之间的摩擦”，根据英文的习惯，前者的位置保持不变，后者则由句中移到了句末。

（2）调查显示，很多夫妻迫于不断加重的经济压力，放弃生育二胎。

The survey shows that many couples have chosen not to have a second child due to the increasing financial stress.

原文中的原因状语因为较长在译文中被放在了句末。

（3）相同学术能力的学生可以同步增加知识。

Students of the same academic ability can increase their knowledge at the same pace.

介宾短语作为状语一般放在句末。

汉语中常会出现表示时间、地点、方式等的多个状语同时出现在一个句子中，汉语的语序为“时间—地点—方式—动词”，英语的语序则为“动词—方式—地点—时间”。请看下面的例子：

（1）大会将于今年九月在北京隆重开幕。

The meeting will begin ceremoniously in Beijing the following September.

如果其中某个状语特别长，也可另作调整。

（2）在这美好的季节，中非友好大家庭的新老朋友们再次相聚，共襄2018年中非合作论坛北京峰会盛举，我们感到十分高兴。

We are so delighted to have all of you with us, friends both old and new, in this lovely season for the reunion of the China-Africa big family at the 2018 Beijing Summit of the Forum on China-Africa Cooperation (FOCAC).

（习近平主席在2018年中非合作论坛北京峰会开幕式上的主旨讲话）

原文中的地点状语很长，所以在译文中简短的时间状语放在了句中，较长的地点状语则放在了句末。使用一系列的时间状语或地点状语时，汉语是按照由大到小的顺序，英语则是遵循由小到大的顺序。

2. 定语位置的调整

汉语中定语通常位于修饰词之前，英语中定语位置分前置和后置两种。如果是单词作定语，一般放在修饰词之前。如果是词组、介词短语或者从句则放在修饰词之后。请看以下例子：

（1）获得有效的治疗。

receive effective treatment

原文和译文的定语位置相同，无需调整。

（2）仅在11月1日，中国消费者就从国内最大的购物平台购买了价值90亿美元的商品。

Only on the November 11th, Chinese consumers bought the goods worth of 9 billion dollars on the biggest shopping platform.

of 词组作为定语，较长，放在修饰词后面。

（3）1996年底，我收到画家黄永玉先生从香港寄来的他个人画展的请柬。

Toward the end of 1996, I received a letter from Mr. Huang Yongyu in Hong Kong, who invited me to his solo art show there.

画线部分的定语处理成了定语从句，放在句末，使句子层次更加分明。

Exercising Your Skills

Translate the following sentences and pay special attention to the order of different parts.

1. 尽管动物实验对于科学研究很重要，但研究人员要减少试验中动物承受的痛苦。

____

2. 你丈夫念书，你挣钱养家，他不会感到不自在吧！

____

3. 丝绸画是一种源于中国，把颜料涂在丝绸布料上的艺术形式，拥有两千多年的历史。

____

4. 一些为接受更好教育而转往城市上学的学生如今又回到了本地农村学校就读。

____

5. 我想回归教育，做我热爱的事情会让我无比兴奋和幸福。

____

Unit Project

Silk is woven into China's economic and artistic history. On the ancient Silk Road, many renowned people left their traces in the history, including eminent diplomats, generals, great monks, merchants, etc. Since 2013, the Belt and Road Initiative (BRI), proposed by Chinese President Xi Jinping, has revived the old trade routes and promotes economic cooperation and cultural communication among Eurasian countries.

On the other hand, in Chinese history, a great amount of people have dedicated themselves to silk craftsmanship and silk industry, including silk embroiders, artists, designers, traders, etc.

Work in groups and report to the whole class a silk-related person that impresses you most. Your report should include the following information:

Unit 7 Keys

1. his / her profile, including name, date of birth, birth place, main personal experiences, etc.;
2. his / her contribution to the silk craftsmanship, silk industry, silk trade, the ancient Silk Road, or the Belt and Road Initiative.

Unit 8 Silk Road

Unit Guide: This is a trade route that began with a single commodity—silk, but changed the world. Extending over tens of thousands of kilometers, the Silk Road ran all the way from China's ancient capital more than 2,000 years ago through the mountains, passes, deserts and oases of Central Asia, and reached the bazaars of Istanbul and the merchants of Venice. People travelling on it not only bought and bartered goods but also exchanged ideas and techniques, and thus the Silk Road cut across borders and brought cultures into contact. In **Passage 1** of this Unit, we will trace through its history to learn that the most important legacy of this ancient trade route is the rich cultural exchange enabled by a time of tolerance. In **Passage 2**, we will have an overview of the Belt and Road Initiative and China's vision of "a global community of common destiny, of shared prosperity, equality and mutual benefit and win-win cooperation."

Lead-in

I Test your knowledge about silk road. Choose the best answer to each of the following sentences.

1. The Silk Road derives its name from the lucrative trade in silk carried out along its length, beginning in the ______ Dynasty.
   A. Qin　　B. Han　　C. Sui　　D. Tang

2. The outstanding diplomat ________ of the Han Dynasty (206 BC—AD 220) traveled the road between 138 BC to 139 BC.
   A. Ban Chao　　B. Ban Zhao　　C. Zhang Qian　　D. Su Wu

3. In the _____ century, when the name of Silk Road was first used by a German geographer, it just included the land road from China's Xinjiang to central Asia.
   A.17th　　B. 18th　　C. 19th　　D. 20th

4. The Silk Road primarily refers to the terrestrial routes connecting East Asia and Southeast Asia with East Africa, West Asia and ________ Europe.
   A. East　　B. South　　C. West　　D. North

5. The first book entitled *The Silk Road* was by______ geographer in 1938.
   A. Swedish B. Swissz C. German D. Chinese

6. Some remnants of what was probably Chinese silk dating from 1070 BC have been found in Ancient ________.
   A. Egypt B. Babylon C. India D. Rome

7. In September 2013, during a visit to _________, Chinese President Xi Jinping introduced a plan for a New Silk Road from China to Europe.
   A. Uzbekistan B. Kazakhstan C. Kyrgyzstan D. Russia

8. In June 2014, UNESCO designated the ___________ corridor of the Silk Road as a World Heritage Site.
   A. Hexi B. Gansu
   C. Gansu-Xinjiang D. Chang'an-Tianshan

9. The maritime parts of the Silk Road involved the following waters EXCEPT__________.
   A. The Yellow River B. The Yellow Sea C. The Persian Gulf D. The Mediterranean

10. The Silk Road involved three continents, namely________________.

A. Europe, America and Asia　　B. America, Africa and Asia

C. Europe, Africa and America　　D. Europe, Africa and Asia

II A brief introduction to the Silk Road.

Step One: Watch the video clip and fill in the blanks with the exact words you hear.

视频

A banker in London sends the latest stocking phone to his 1.________ in Hong Kong in less than a second. With a single click, a customer in New York orders 2.________ in Beijing transported across the ocean within days by cargo plane or 3.________. The speed in volume which goods and information move across the world today is 4.________ in history. But global exchange itself is older than we think. Reaching back over two thousand years along a five thousand miles 5.________ —known as the Silk Road. The Silk Road wasn't actually a 6.________ road, but in network of 7.________ routes that gradually emerged over centuries connecting various settlements into each other 8.________.

The first agricultural civilizations were isolated places in 9.________ river valleys. Their travel 10.________ by surrounding geography and fear they (were) unknown.

Step Two: Watch the video clip again and decide whether the following statements are true or false. Write T for true and F for false.

1. ( ) It's regular relay points allows goods and messages to travel in nearly one fifth the time that would take a single traveler.

2. ( ) It was realized in the first century BC when an ambassador named Zhang Qian sent a negotiate with Roman in the west retuned to the Henan Province.

3. (   ) Chinese goods make their way to Rome causing an outflow of gold that led an abandon on silk while Roman glasswork was highly priced in China.

4. (   ) Originating in India, Buddhism migrated into China and Japan to become the dominated religion there.

5. (   ) The collapse of Rome rule was followed by China's withdraw from international trade.

6. (   ) Today, global interconnectedness shapes our lives like before.

7. (   ) The impact of globalization on culture and economy is disputable.

8. (   ) None of it would have been possible without the pioneering cultures whose efforts create the Silk Road, history's first worldwide web.

Ⅲ Topic discussion.

1. What's the significance of the Silk Road? Can you explain from the aspects of history, economy, culture or tourism?

2. It's said that only the brave men can cross the Silk Road. Can you choose one of the historic figures who crossed the Silk Road and make a profile for him?

Passage 1

The Legacy of the Silk Road

Valerie Hansen[1]

NEW HAVEN: Despite all the talk in diplomatic circles of a new Silk Road and restoring trade in Central Asia, in actuality, these routes were among the least traveled in human history—possibly not worth studying if tonnage, traffic or the number of travelers at any one time were sole measures. The Silk Road found a place in history because of its rich cultural legacy in written records and artifacts, and because trade and tolerance were so **intertwined**[2].

Trade was not the primary purpose of the Silk Road, more a network of pathways than a road, **in its heyday**. Instead, the Silk Road changed history, largely because the people who managed to travel along part or all of the Silk Road planted their cultures like seeds of exotic species carried to distant lands. Thriving in new homes, newcomers mixed with local residents and often absorbed other groups who followed. Sites of sustained economic activity, oasis towns like **Turfan**, **Dunhuang** or **Khotan**, **enticed** still others to cross over mountains and **traverse** oceans of sand. While not much of a commercial route, the Silk Road became the planet's most famous cultural **artery** for the exchange between East and West of religions, art, languages and new technologies.

We use the term "Silk Road" to refer generally to the exchanges between China and places farther to the west, specifically Iran, India and, on rare occasions, Europe. Most **vigorous** before the year 1000, these exchanges were often linked to Buddhism.[3]

And that's why cities of Khotan and Kashgar in Xinjiang, northwestern China, are famous for their Sunday markets, where tourists can buy locally made crafts, **naan** and grilled mutton on **skewers**. As visitors watch farmers fiercely bargaining over the price of a donkey, it's easy to imagine Xinjiang always this way, but that's an illusion. The **predominantly** non-Chinese crowds in the northwest **prompt** a similar reaction: Surely these are the direct descendants of the earliest Silk Road settlers.

In fact, though, a major historic break divides modern Xinjiang from its Silk Road past. The **Islamic** conquest of the Buddhist kingdom in 1006 brought a dramatic **realignment** to the region. Eventually Xinjiang's inhabitants converted to **Islam** making that the principal religion in the region today. They also gradually gave up speaking **Khotanese**, **Tocharian**, **Gandhari** and other languages spoken during the first millennium AD for **Uighur**, the language one hears most often in the region today.

Excavated materials **shed light on** the nature of the Silk Road trade. These materials, written on paper, silk, leather and wood, survive only in dry locales, places like **Niya, Loulan**, **Kucha**, Turfan and Khotan in Xinjiang; **Samarkand** in **Uzbekistan**; Chang'an, Dunhuang in Gansu province; and Chang'an, the capital during the Former Han dynasty (206 BC—AD 9) and the Tang (618—907). These documents were recovered not only from tombs, but also from abandoned postal stations, **shrines** and homes, beneath the dry desert—the perfect environment for the preservation of documents as well as art, clothing, ancient religious texts, **ossified** food and human remains.

Many documents, found by accident, were written by people from all social levels, not simply the literate rich and powerful. These documents were not composed as histories. Their authors did not expect later generations to read them, yet they offer a glimpse into the past that's often **refreshingly** personal, factual, **anecdotal**, and random.[4]

Documents later recycled as shoes for the dead or in the arms of **figurines** show that Silk Road trade was often local and small in scale. Even the most **ardent** believer in a high-volume, frequent trade must **concede** that there is little **empirical** basis. Scholars offer varying interpretations of these scraps of evidence, but there's no denying that the debates concern scraps, not massive bodies, of evidence.

The modern discovery of the Silk Road began in 1895 when the Swedish explorer Sven Hedin launched his first expedition into the **Taklamakan Desert** in search of the source of the Khotan River. After 15 days, he discovered that he was not carrying enough water for himself and the four men with him. He did not turn back, not wanting to admit his expedition had failed. When their supply ran out, he began a desperate search, eventually locating a stream, but not before two men **perished**.

As he made his way out of the desert, Hedin encountered a **caravan** of merchants and pack animals, and he purchased three horses, **saddles**, maize, flour, tea, utensils and boots. This list, described in his biography, is revealing. Even at the beginning of the 20th century, almost all the goods traded in the Taklamakan were locally made necessities, not foreign imports.

Similarly, during the first millennium, markets offered more local goods for sale than foreign-made imports. At one market in Turfan in 743, local officials recorded prices for 350 items, including typical Silk Road goods like **ammonium chloride**, used for dyeing cloth and softening leather, as well as **aromatics**, sugar and brass. Of course, locally grown vegetables, staples and animals, some brought over long distances, were also available.

Despite the limited trade, cultural exchange between East and West was extensive—first between China and South Asia, and later west Asia, especially Iran. Refugees, artists, craftsmen, missionaries, robbers and envoys traveled along these routes in Central Asia. The most influential people moving along the Silk Road were refugees. Waves of immigrants brought technologies from their respective homelands, practicing those skills or introducing **motifs** in their new homes.

Frequent migrations of people fleeing war or political conflicts meant that some technologies moved east, others west. As techniques for making glass entered China from the Islamic world, the technology for manufacturing paper was transported westward. Invented in China during the 3rd century BC, paper moved out of China, first to **Samarkand**, arriving sometime around the year 700, and then into Europe from the Islamic **portals** of **Sicily** and Spain. Paper, the most convenient and affordable material for preserving writing, encouraged great cultural change, including the printing revolution in Western Europe. Of course, the Chinese developed woodblock printing much earlier than **Gutenberg**, starting around 700.

Cultural transfer took place as the Chinese learned from other societies, specifically India, the home of Buddhism. Buddhist missionaries were key translators and worked out a system for transcribing unfamiliar terms in foreign languages, like **Sanskrit**, into Chinese that remains in use today. Chinese absorbed some 35,000 new words, including both technical Buddhist terms and common everyday words.

People who spoke different languages often encountered one another on the Silk Road. Some had learned multiple languages since childhood. Others had to learn foreign languages as adults, a more arduous process than it is today given how few study aids were available. Surviving phrasebooks shed light on student identities and reasons for their studies.[5] Used in monasteries throughout the first millennium, Sanskrit attracted students, but so, too, did Khotanese, Chinese, and Tibetan.

The most important legacy of the Silk Road is the atmosphere of tolerance fostered by rulers of small oasis kingdoms strung along the northern and southern Taklamakan. Over the centuries these rulers welcomed refugees from foreign lands, granting them permission to practice their own faiths. Buddhism entered China, and so too did **Manicheism**, **Zoroastrianism** and the **Christianity** of the East.[6] Archeological sites and the preserved artifacts offer a glimpse into this once tolerant world. The new Silk Road is indeed far removed from the legacy of the historic network.

Words

intertwine [ɪntə'twaɪn] *vt.& vi.* 缠绕；纠缠
entice [ɪn'taɪs] *vt.* 诱使；怂恿
traverse [trə'vɜːs] *vt.* 穿过；反对
artery ['ɑːtəri] *n.* 动脉；干道；主流
vigorous ['vɪgərəs] *adj.* 有力的；精力充沛的
nann [nɑːn] *n.* 馕；印度烤饼；印度薄饼
skewer ['skjuːə] *n.* 烤肉叉子；串肉扦；针
predominantly [prɪ'dɒmɪnəntli] *adv.* 占主导地位地；显著地；占优势地
prompt [prɒmpt] *vt.* 提示；促进；激起
Islamic [ɪz'læmik] *adj.* 伊斯兰教的
realignment [ˌriːə'laɪnmənt] *n.* 重新排列；重新组合；改组
Islam ['ɪzlɑːm] *n.* 伊斯兰教；伊斯兰教徒；伊斯兰教国家
excavate ['ekskəveɪt] *vt.& vi.* 挖掘；开凿
shrine [ʃraɪn] *n.* 圣地；神殿；神龛
ossified ['ɔsɪfaɪd] *adj.* 僵化的；已骨化的

refreshingly [rɪˈfreʃɪŋli] *adv.* 清爽地；有精神地
anecdotal [ˌænɪkˈdəʊt(ə)l] *adj.* 轶事的；趣闻的；多轶事的；含轶事的
figurine [ˈfɪgəriːn] *n.* 小雕像；小塑像
ardent [ˈɑːdnt] *adj.* 热情的；热心的；激烈的；燃烧般的
concede [kənˈsiːd] *vt.* 承认；退让；给予，容许
empirical [imˈpɪrɪkl] *adj.* 经验主义的；凭经验的
perish [ˈperɪʃ] *vi.* 死亡；毁灭
caravan [ˈkærəvæn] *n.* （可供居住的）拖车；大篷车；（穿过沙漠地带的）旅行队（如商队）
saddle [ˈsædl] *n.* 鞍；鞍状物；车座；拖具
ammonium [əˈməʊniəm] *n.* 铵；氨盐基
chloride [ˈklɔːraɪd] *n.* 氯化物
aromatic [ˌærəˈmætɪk] *n.* 芳香植物；芳香剂；香料
motif [məʊˈtiːf] *n.* 主题；动机；主旨；意念
portal [ˈpɔːtl] *n.* 入口；桥门
Sanskrit [ˈsænskrɪt] *n.* 梵文；梵语

Useful Expressions

in its heyday 在全盛时期；在繁荣时期
shed light on 为……提供线索；对……透露情况；使……清楚地显出

Proper Names

New Haven 纽黑文（美国地名）
Turfan 吐鲁番（新疆地名）
Dunhuang 敦煌（甘肃地名）
Khotan 和田（新疆地名）
Khotanese 和田语
Tocharian 吐火罗人；吐火罗语
Gandhari 犍陀罗语
Uighur 维吾尔语；维吾尔人；维吾尔族
Niva 尼雅遗址（今新疆境内）
Loulan 楼兰（西域古国遗址）
Kucha 库车，古称龟兹（古代西域大国）

Samarkand　撒马尔罕（乌兹别克斯坦东部城市）
Uzbekistan　乌兹别克斯坦
Taklamakan Desert　塔克拉玛干沙漠
Sicily　西西里岛（意大利地名）
Gutenberg　古腾堡（Johannes，1400—1468，德国活版印刷发明人）
Manicheism　摩尼教
Zoroastrianism　拜火教；索罗亚斯德教；明教
Christianity　基督教

Notes

1. Valerie Hansen teaches Chinese history at Yale University. Her most recent book is *The Silk Road: A New History* (Oxford University Press, 2012).
瓦莱丽·汉森在耶鲁大学教授中国历史。她最近的一本书是《丝绸之路：新历史》（牛津大学出版社，2012年）。

2. The Silk Road found a place in history because of its rich cultural legacy in written records and artifacts, and because trade and tolerance were so intertwined.
丝绸之路在历史上占有一席之地，因为它以书面记录留存下来的丰富的文化遗产以及手工艺品，也因为贸易和包容在此共存。

3. We use the term "Silk Road" to refer generally to the exchanges between China and places farther to the west, specifically Iran, India and, on rare occasions, Europe. Most vigorous before the year 1000, these exchanges were often linked to Buddhism.
我们一般用"丝绸之路"这一术语，代指中国和遥远的西方国家，尤其是与伊朗、印度，偶尔与欧洲之间的物物交换。这些交易在公元1000年之前最活跃，通常与佛教相关。

4. Many documents, found by accident, were written by people from all social levels, not simply the literate rich and powerful. These documents were not composed as histories. Their authors did not expect later generations to read them, yet they offer a glimpse into the past that's often refreshingly personal, factual, anecdotal, and random.
很多偶然发现的文件由社会各阶层的人们书写，不仅仅是有钱有势有文化的人群。这些文件并非作为历史来记录，它们的作者也并不指望后人能读到它们，但使人们得以一窥这些令人耳目一新的有关个人的、充满逸闻趣事的、真实又随意的往事。

5. People who spoke different languages often encountered one another on the Silk Road. Some had learned multiple languages since childhood. Others had to learn foreign languages as adults, a more arduous process than it is today given how few study aids were available. Surviving phrasebooks shed light on student identities and reasons for their studies.
说着不同语言的人们常常能在丝绸之路上遇见彼此。有些人从童年开始就学了多种语言，还有些人成年以后才不得不学外语。当时没有什么学习工具，比现在的学习艰难得多。残存下来的词汇手册清楚地显示了学生的身份以及他们学习的原因。

6. The most important legacy of the Silk Road is the atmosphere of tolerance fostered by rulers of small oasis kingdoms strung along the northern and southern Taklamakan. Over the centuries these rulers welcomed refugees from foreign lands, granting them permission to practice their own faiths. Buddhism entered China, and so too did Manicheism, Zoroastrianism and the Christianity of the East.
丝绸之路最重要的遗产是塔克拉玛干沙漠南北沿线的一些绿洲王国的统治者们所扶持的一种包容的氛围。几个世纪以来，这些统治者们欣然接受国外的难民，允许他们有自己的信仰。佛教传入中国，随之而来的还有摩尼教、拜火教以及东方基督教。

Reading Comprehension

Answer the following questions according to Passage 1.

1. Why did the Silk Road find a place in history?

2. What do we usually use the term "Silk Road" to refer to?

3. What did the Islamic conquest of the Buddhist kingdom in 1006 bring about?

4. Who wrote the documents found by accident?

5. When and what marked the beginning of the modern discovery of the Silk Road?

6. Who were the most influential people moving along the Silk Road and how?

7. Who were key translators and worked out a system for transcribing unfamiliar terms in foreign languages?

8. What is the most important legacy of the Silk Road?

Summary

The following is a summary of Passage 1. Fill in the blanks with appropriate words.

The Silk Road, an ancient 1.________ of routes crisscrossing the continent for trade and security. But Valerie Hansen, author and professor of history at Yale University points out that 2._______ was not the primary purpose of the network. "Instead, the Silk Road changed 3.________, largely because the people who managed to travel along part or all of the Silk Road planted their cultures like 4.________ of exotic species carried to distant lands," she writes "Thriving in new homes, newcomers mixed with local residents and often absorbed other groups who followed." The cultural exchanges were rich and in many cases lasting, as suggested by 5.________ materials and documents, prepared by people of all social 6._______ centuries ago and preserved in the 7.______ climates. Silk Road traffic may not have been heavy, but cultural exchange was extensive and rich during an era of 8.________.

Word Building

I The prefix *out* can be used together with other words to form new words, meaning "to go beyond, surpass, exceed, remove." Match the following words with their parts of speech and Chinese meanings.

	outdo	突发
	outlive	吃得比……多
	outeat	比……长命
n.	outnumber	剥夺法律权利
	outlaw	超过数目
adj.	out-party	户外的
	outsize	特大号
v.	outbreak	在野党
	outroot	除根
	outdoor	胜过；超越

II The prefix *over* can be used together with other words to form new words, meaning "to be excessive, upside, upend." Match the following words with their parts of speech and Chinese meanings.

	overbridge	极度高兴的
	overseas	饮酒过量
n.	overcoat	过奖
	overpraise	俯瞰
adj.	overpay	超载；负荷
	overdrink	付得过多
v.	overload	海外的
	overlook	外套
	overjoyed	天桥
	overturn	颠覆

Words and Expressions in Use

I Fill in the blanks with the words or expressions given below. Change the form where necessary. Each word or expression can be used only once.

intertwine	in its heyday	artery	vigorous	predominantly	prompt
shed light on	refreshingly	anecdotal	ardent	concede	perish

1. ____________, the silk market received 10,000 visitors one day, the manager recalled.

2. The scenes on stage shift between two time frames as the live of Zhu and Hamlet unfold and to some degree ______.

3. A large ______ carrying the nation's cultural genes, the canal belongs to the old days, but it is also for today and the future.

4. My grandmother looks _______ in the picture, but now she has Alzheimer's.

5. Now more and more youngsters are falling in love with hanfu, which name is given to pre-17th century traditional clothing of the Han Chinese, the country's ________ ethnic group.

6. The president said he would work with Putin to _________ bilateral relations and

cooperation in various fields to make new progress and bring more benefits to the two countries and people.

7. China launches Chang'e-4 probe to ___________ moon's dark side.

8. His debut album offers a ________ eclectic mix of musical styles that makes it some ideal jazz listening for a summer's day.

9. An exhibition at the National Museum of Classic Books in Beijing unveils the _______ behind *The Great Canon of the Yongle Era.*

10. What has impressed him most have been the ________ aspirations of people around the world for peace, tranquility, development, progress and a happier life.

11. "We didn't ________ a goal in six games, it's one of the main keys why we qualified for the final," said Qatar cocah Felix Sanchez.

12. Among academics, the maxim "publish or ______" (i.e., publish your research or risk losing your job is a threatening reminder of the importance of publication.

II Render the following Chinese expressions into English. The first letter of each word is given.

1. 丰富的文化遗产 r__________ c___________ l________
2. 文化主干线 c___________ a___________
3. 戏剧性重组 d____________ r____________
4. 出土材料 e___________ m____________
5. 社会各阶层 a____________ s____________ l________
6. 狂热的信徒 a____________ b____________
7. 塔克拉玛干沙漠 T____________ D____________
8. 佛教术语 B____________ t____________
9. 绿洲王国 o____________ k____________
10. 可见一斑 o_________ a___________ g_________

Translation

Translate the following paragraph into English.

丝绸之路是历史上连接中国和地中海的一条重要贸易路线。因为这条路上的丝绸贸易占多数，故而在1877年被德国的一位杰出的地理学家命名为“丝绸之路”。同时，丝绸之路也是许多其他商品进行交易的主干线。更重要的是，也是传播技术的一条主要通道。通过丝绸之路传入中国的事物中意义最为重大的当属佛教。中国的养蚕技术、生铁锻造技术和灌溉技术曾通过丝绸之路传播到中亚、南亚和欧洲。

Passage 2

The New Silk Road—OBOR—an Overview[1]

I have been following China, the events there and the development of the country, all my life. It has been the greatest privilege of my life to be a witness to and, if I may say so, a modest participant in, one of the greatest chapters in human history. I refer to the remarkable transformation and development of China in an unprecedentedly short period of time into the world's second largest economy and the transition from poverty to the beginnings of **moderate** prosperity that this has **entailed** for well over a billion people. Given the **majesty** of China's history, culture and civilisation, I prefer to refer to this not as the rise of China but rather as the return of China.

Now that China has returned, the question of what sort of country it will be, what role it will play in the world, is something that naturally concerns the entire international community.

It is against this background that China has advanced the concepts of a global community of common destiny, of shared prosperity, equality and mutual benefit and win-win cooperation. China, having suffered from poverty and **oppression** in the past, understands that a world where only China develops, and its neighbours and the wider world either **stagnate** or **relapse into** poverty and conflict, is not only undesirable, but also actually impossible. China itself has prospered by opening its door to the outside world. Sustainable and lasting prosperity has to mean prosperity for all.

This is the thinking behind a whole range of China's recent initiatives—for example, of the development of BRICS[2], along with Brazil, Russia, India and South Africa, and the creation of their New Development Bank; of the **Asian Infrastructure Investment Bank (AIIB)**, formally launched in Beijing earlier this month; and especially of our topic of the moment, the great vision of jointly building the Silk Road Economic Belt and the 21st-Century Maritime Silk Road.

As Minister Zhang has already explained, this is a new idea, but one with an ancient **lineage**. As the **National Development and Reform Commission**, the Ministry of Foreign Affairs and the Ministry of Commerce of the People's Republic of China have jointly **observed**:

"More than two millennia ago the diligent and courageous people of Eurasia explored and opened up several routes of trade and cultural exchanges that linked the major civilisations of Asia, Europe and Africa, collectively called the Silk Road by later generations. For thousands of years, the Silk Road Spirit—'peace and cooperation, openness and inclusiveness, mutual learning and mutual benefit'—has been passed from generation to generation, promoted the progress of human civilisation, and contributed greatly to the prosperity and development of the countries along the Silk Road. Symbolising communication and cooperation between the East and the West, the Silk Road Spirit is a historic and cultural heritage shared by all countries around the world."

China's initiative to jointly build the Belt and Road, embracing the trend towards a **multipolar** world, economic globalisation, cultural diversity and greater IT application, is designed to uphold global free trade and an open world economy and to enhance regional cooperation. It aims at being highly efficient in terms of the **allocation** of resources and at achieving a deep integration of markets among the countries along the Belt and Road, thereby jointly creating an open, inclusive and balanced regional economic cooperation architecture that benefits all.

According to the vision of the Chinese government, the Belt and Road Initiative[3] is in line with the purposes and principles of the United Nations **Charter**[4]. It upholds the Five Principles of Peaceful Coexistence[5], namely mutual respect for each other's **sovereignty** and territorial integrity, mutual non-aggression, mutual non-interference in each other's internal affairs, equality and mutual benefit, and peaceful coexistence.

The initiative is an open one. It covers, but is not limited to, the area of the ancient Silk Road. It is open to all countries, and international and regional organisations, so that the results will benefit wider parts of the globe as well.

It is harmonious and inclusive. It advocates tolerance among civilisations, respects the paths of development chosen by different countries, and supports dialogues among different civilisations on the principles of seeking common ground whilst reserving differences and **drawing on** each other's strengths, so that all countries can coexist in peace for common prosperity.

The New Silk Road follows market principles. It will abide by market rules and international norms, will give play to the decisive role of the market in resource allocation and the primary role of enterprises, and will also let governments perform their due functions.

The New Silk Road is **envisaged** to go in five directions:

—From Northwest and Northeast China through Central Asia and Russia to the **Baltic Sea**;

—From Northwest China through Central Asia and the Persian Gulf to the Mediterranean;

—From Southwest China through the Indochina peninsula, Malaysia and Singapore to the Indian Ocean;

—From the Chinese ports, through the South China Sea and the **Straits of Malacca** to the Indian Ocean and westwards from there, for example to East Africa;

—And by the same route but then on to the South Pacific from the Straits of Malacca.

Six economic **corridors** are envisaged:

—From Northeast China through Mongolia and Russia to **the Baltics**;

—From the coastal provinces through western China to Central Asia and then to Russia and the Baltics;

—From Northwest China through Xinjiang, Central and West Asia to the Persian Gulf and the Mediterranean;

—From Yunnan and Guangxi Zhuang in Southwest China through Vietnam. Laos, Cambodia, Thailand and Malaysia to Singapore;

—From China through Pakistan, entering the Indian Ocean through the port of **Gwadar**;

—And, through Myanmar, **Bangladesh** and India, entering the Indian Ocean via the **Bay of Bengal**.

These new silk routes will embrace—and will require major investments in—railways, highways, sea transportation, pipelines and the information superhighway and **connectivity**.

To translate this grand vision into reality will require trillions of dollars of investment in **infrastructure** and in all sectors of the economy in the more than 60 countries directly **encompassed** within the new silk road initiative, as well as further **afield**. **Cumulatively** it represents the greatest business opportunity in the contemporary world.

One Belt One Road may be the official name given to this grand project, but personally I prefer the New Silk Road. My own study of history has taught me that President Xi's idea of the New Silk Road takes the past and updates it brilliantly, with a view to managing both present and future challenges.

However, the progress of science, technology and civilisation will surely sweep away many of the obstacles, and hence the need to make some of the **detours** that were required in the past. For example, a future high-speed train running from Urumqi to Istanbul will use new technologies, quite likely including magnetic **levitation**, to shorten the route and create a new way.

Across three great continents, the new silk roads will create new ways by using the three highs, namely:

High speed trains;

High speed energy **transmission**;

High speed connectivity and communications.

...

The future promise of the New Silk Road will, therefore, not be attained easily and the difficulties and obstacles should never be under-estimated. Yet it is not naive to **hold out** this vision for the future. Viewed correctly, it might rather be said to be a supreme act of realism. For without hope, without development, without knowledge, without prosperity, how can there ever be lasting peace; how can hatred ultimately make way for coexistence, mutual respect and **amity**?

Rather this project will work precisely because it will be based upon the full participation of all national governments, their **procurement** agencies, and on global competition based on market principles, governed by the rule of law and subject to transparent **tendering**, **oversight**, regulation and **arbitration**.

We shall also play a significant part in legal, accounting, financial, regulatory and other professional services as well as in funding and listings.

This is a One Belt One Road for all—the ultimate win-win.

I believe that it will change the world forever and leave behind empires and **hegemony** in favour of a community of common destiny, where the development of one nation is the **precondition** for the development of all nations. Back to the future. **Ecumene** shall return.

...

(The article is authorised to People's Daily Online by Stephen Perry, who gave this speech on China Development Forum at the L.S.E. 30 January 2016.)

Words

moderate [ˈmɒdəreɪt] *adj.* 有节制的；稳健的；适度的，中等的

entail [ɪnˈteɪl] *vt.* 需要；使必要

majesty [ˈmædʒəsti] *n.* [U] 雄伟壮观；庄严；威严

oppression [əˈpreʃn] *n.* 压迫；被压迫的状态

stagnate [stægˈneɪt] *vi.* 停滞；不流动；不发展；变萧条

relapse [rɪˈlæps] *vi.* 故态复萌；复发；再度堕落

lineage [ˈlɪniɪdʒ] *n.* 世系；宗系；家系；血统

observe [əbˈzəːv] *vi.* 注意；说；评述

multipolar [ˌmʌltɪˈpəʊlə] *adj.* 多极的

allocation [ˌæləˈkeɪʃn] *n.* 配给；分配；分配额（或量）；划拨的款项；拨给的场地

charter [ˈtʃɑːtə(r)] *n.* 宪章

sovereignty [ˈsɒvrɪnti] *n.* 主权国家；国家的主权

envisage [ɪnˈvɪzɪdʒ] *vt.* 想像，设想；观察，展望

strait [streɪt] *n.* 海峡

corridor [ˈkɒrɪdɔː(r)] *n.* 走廊，通道；走廊（一国领土通过他国境内的狭长地带）

connectivity [ˌkɒnekˈtɪvəti] *n.* 连通性

infrastructure [ˈɪnfrəstrʌktʃə] *n.* 基础设施；基础建设

encompass [ɪnˈkʌmpəs] *vt.* 包含或包括某事物

afield [əˈfiːld] *adv.* 在战场上；在野外；远离着

cumulatively [ˈkjuːmjʊlətivlɪ] *adv.* 累积地；渐增地

detour [ˈdiːtʊə(r)] *n.* 绕道；弯路

levitation [ˌlevɪˈteɪʃ(ə)n] *n.* 升空；漂浮；浮起

transmission [trænsˈmɪʃn] *n.* 传送；传动装置

amity [ˈæməti] *n.* 和睦，友好关系

procurement [prəˈkjʊəmənt] *n.* 采购；获得；取得

tender [ˈtendə(r)] *v.* 投标；（正式）提出

oversight [ˈəʊvəsaɪt] *n.* 监督；照管

arbitration [ˌɑːbɪˈtreɪʃn] *n.* 仲裁；公断

hegemony [heˈdʒəməni] *n.* 霸权；霸权主义

precondition [ˌpriːkənˈdɪʃn] *n.* 前提；先决条件

ecumene [ˈekjʊmiːn] *n.* （适于人类）居住的部分；永久栖居地

Useful Phrases

relapse into 复发；退回原处
draw on / upon 利用；凭借
hold out 伸出；拿出；呈现；抵抗

Proper Names

Asian Infrastructure Investment Bank (AIIB) 亚洲基础设施投资银行
National Development and Reform Commission 国家发展和改革委员会
Baltic Sea （地名）波罗的海
Straits of Malacca （地名）马六甲海峡
the Baltics 波罗的海国家
Gwadar （地名）瓜达尔港
Bangladesh （国家名）孟加拉国
Bay of Bengal （地名）孟加拉湾

Notes

1. This passage was adapted from the speech of Stephen Perry on China Development Forum at the LSE (London School of Economics and Politics) on 30 January 2016 and published by People's Daily Online on 5 February, 2016.
本文改编自斯蒂芬•佩里于2016年1月30日在伦敦政治经济学院论坛上关于中国发展的讲话，人民日报网2016年2月5日刊发。

2. BRICS is the acronym coined for an association of five major emerging national economies: Brazil, Russia, India, China and South Africa. Originally the first four were grouped as "BRIC" (or "the BRICs"), before the induction of South Africa in 2010. The BRICS members are known for their significant influence on regional affairs; all are members of G20. Since 2009, the BRICS nations have met annually at formal summits.
金砖国家（BRICS），特指世界新兴市场，因其引用了巴西（Brazil）、俄罗斯(Russia)、印度(India)、中国（China）、和南非（South Africa）的英文首字母。由于该词与英语单词的砖（brick）类似，因此被称为"金砖国家"。金砖国家同时也是20国集团的成员，自2009年以来，金砖国家领导人每年举行一次正式会晤。

3. The Belt and Road Initiative (BRI), also known as the One Belt One Road (OBOR) or the Silk Road Economic Belt and the 21st-century Maritime Silk Road, is a development strategy adopted by the Chinese government involving infrastructure development and investments in countries in Europe, Asia and Africa.
   “一带一路”即“丝绸之路经济带和21世纪海上丝绸之路”的简称，“一带一路”的英文表达是The Belt and Road Initiative或One Belt One Road，英文缩写是BRI或OBOR。“一带一路”是由中国政府提出的一项发展倡议，深化与亚洲、非洲和欧洲国家的经济合作，体现同舟共济、权责共担的命运共同体意识，完善全球治理体系变革的新思路新方案。

4. The United Nations Charter: The Charter of the United Nations (also known as the UN Charter) of 1945 is the foundational treaty of the United Nations. The UN Charter outlined a broad set of principles relating to achieving “higher standards of living,” addressing “economic, social, health, and related problems,” and “universal respect for, and observance of, human rights and fundamental freedoms for all without distinction as to race, sex, language, or religion.” As a charter, it is a constituent treaty, and all members are bound by its articles.
   《联合国宪章》于1945年通过，是联合国的基本大法，它既确立了联合国的宗旨、原则和组织机构设置，又规定了成员国的责任、权利和义务，以及处理国际关系、维护世界和平与安全的基本原则和方法。

5. Five Principles of Peaceful Coexistence, known as the Panchsheel Treaty: Non-interference in others internal affairs and respect for each other’s territorial unity integrity and sovereignty are a set of principles to govern relations between states. Their first formal codification in treaty form was in an agreement between China and India in December 1953. They were enunciated in the preamble to the “Agreement (with exchange of notes) on Trade and Intercourse Between Tibet Region of China and India,” which was signed at Beijing on 29 April 1954.
   和平共处五项原则也被称为潘基塞条约。这五项原则最先是周恩来总理于1953年12月底在会见来访的印度代表团时提出的，当时中国政府同印度政府就两国在西藏地方的关系问题进行谈判，这五项原则即“互相尊重主权和领土完整，互不侵犯，互不干涉内政，平等互利，和平共处”。1954年4月29日，中印两国发表谈判公报，并签署了《中印关于中国西藏地方和印度之间的通商和交通协定》，两国政府一致同意把和平共处五项原则列入公报和协定中，把它作为指导两国关系的准则。

Reading Comprehension

Answer the following questions according to what you read from Passage 2.

1. What do you learn about the author from this passage?

2. What is this passage mainly about?

3. Why does the author refer to China's transformation and development as the return of China instead of the rise of China?

4. Why did China advance the concept of "a global community of common destiny"?

5. Why is Belt and Road Initiative proposed by China a new idea with an ancient lineage?

6. What are the Five Principles of Peaceful Coexistence?

7. In what ways are Belt and Road Initiative an open one?

8. What will the new Silk Road be like if translated into reality?

9. Why does the author prefer to call the grand project "the New Silk Road" rather than "One Belt and One Road"?

10. What does the author believe to be the future of Belt and Road Initiative?

Words and Experssions in Use

I Match the collocations in Column A with the Chinese meanings in Column B.

Column A	Column B
A. Chinese initiative	1. 求同存异
B. a community of common destiny	2. 一带一路
C. Asian Infrastructure Investment Bank	3. 中国倡议
D. seek common ground whilst reserving differences	4. 经济走廊
E. cargo trains	5. 国际仲裁
F. One Belt One Road	6. 命运共同体
G. magnetic levitation	7. 亚洲基础设施投资银行
H. BRICS	8. 货运火车
I. international arbitration	9. 磁悬浮
J. economic corridor	10. 金砖国家

II Fill in the blanks with the words or expressions given below. Change the form where necessary. Each word or expression can be used only once.

moderate	stagnate	lineage	multipolar	encompass	procurement
tender	envisage	entail	relapse into	draw on	hold out

1. Fortunately, the financial world is a far more liberal, ___________ place than it used to be.

2. He also thinks GDP will likely ___________ negative territory in the third quarter, given the problems still dogging the economy.

3. Real wages rely directly on investment; without increases in productivity, wages are bound to ___________ at best.

4. On leaving office, they would also be banned for at least a year from joining any company ___________ for public contracts.

5. A few laboratory studies even suggest that ___________ caffeine intake may help protect the heart against this problematic rhythm disorder.

6. If the BRICs pursue sound policies, the world we___________ here might turn out to be a reality, not just a dream.

7. The Arab uprisings ___________ the long-term prospect of a more stable and prosperous neighbourhood south of the Mediterranean.

8. To merit the name, World Heritage sites need to ___________ the intangibles, to be virtual at least as much as they're physical.

9. The WTO rulebook needs to be updated and strengthened with tougher rules against government subsidies, intellectual property rights protections, and rules covering public ___________ and liberalization of services.

10. Although you are focusing on a single moment in the text, you may ___________ what you know about the author and his or her choices elsewhere.

11. Cleopatra Ⅶ was born in Egypt, but she was descended from a ___________ of Greek kings and queens who had ruled Egypt for nearly 300 years.

12. Staff training is one of the chief ways to enhance staff performance. These will ____________ practical, effective and targeted training.

Ⅲ Complete the following statements by translating the Chinese in brackets into English.

1. One Belt One Road is famous ______________________（因为它是中国的倡议）, and certainly Chinese involvement is the key in making this plan.

2. ____（参与新丝绸之路及相关项目的国家、企业、发展银行）are also international, and span across the entire Eurasian theater.

3. Since China proposed the BRI in 2013, more than 140 countries and international organizations including Portugal ____________________________________（签署了合作协议）under the initiative.

4. Xi'an, known in ancient times as Chang'an, is the starting point of the Silk Road and ___ ____（在一带一路中起着重要的作用）.

5. Following the commencement of the initiative, ______________________________（一带一路沿线经济体的平均得分超过了）the non-Belt and Road region, and the gap is widening.

6. The Belt and Road Initiative is expected to strengthen connectivity, boost economies and raise people's living standards ____（在参与一带一路的国家和地区）, analysts from different countries said.

7. The UNESCO-initiated Silk Road Project gives value to the historic legacy, __________ ____（把东西方融合起来的文化和思想交流）thanks to this route.

8. The Belt and Road Initiative is not a repeat of the Marshall Plan but about ___________ ____（分享中国的经济经验或中国的解决办法）.

Translation Skills

句子英译（4）——否定句的翻译

英汉两种语言中都存在否定句，共同之处在于都会采用表示否定意义的词来传达否定概念。不同之处在于英语除了采用词汇手段以外，还会采用句法手段，来表达否定意义，而且英语的词汇手段要丰富得多，可以是no、not、never、nor、neither等否定词，或者是由表示否定意义的词缀构成的词，也可以是表示否定意义的副词、名词、动词、介词和连词等。

汉语否定句的英译主要分以下几种情况。

1. 译成英语的否定句

汉语的否定句转换为英语否定句时，请特别注意否定词的位置。例如：

（1）没有听到声音。

No sound was heard.

原句中否定词放在动词前，译文中把原来的宾语转化成了主语，否定词也放在主语前。

（2）我觉得他们不会反对我的建议。

I don't think they will object to my suggestions.

英语中否定词一般放在主句的谓语动词前，而不是从句的谓语动词前。

否定句英译时需注意完全否定和部分否定的区别。

（3）并不是每个人都相信谣言。

Not everybody believes in rumors.

（4）我的朋友都不抽烟。

None of my friends smoke.

如句（3）所示，否定词not与all、both、everyone、everything等不定代词一起使用时表示部分否定。要是完全否定，必须像句（4）使用全否定词，如none、neither、no、nobody、nothing等。

2.译成英语的肯定句

第一种：形式肯定，意义也肯定。

汉语和英语两种语言有时对相同的概念，表达的角度刚好相反。例如:

（1）欲速则不达。

Haste makes waste.

（2）这个公园美得无法形容。

The beauty of the park is more than words can describe.

上文提过英语可以采用句法手段来表达否定意义，比较结构就是其中的一种，常见的如other than..., rather than..., would rather than..., know better than..., might as well... than...等。

（3）与其把书给他，你还不如把书烧了。

You might as well burn these books than give them to him.

第二种：形式肯定，意义否定。

（1）那座木桥一点也不安全。

That wood bridge is anything but safe.

（2）我没有学那种技术。

I have yet to learn the skill.

两句译文中的anything but和have yet to do形式虽然肯定，意义却是否定。

（3）他说他要是放手就不是人。

He said he would be damned if he'd stop.

这句译文通过if引导的句型表示诅咒、发誓的语气，从而来传达否定的意义。

3. 汉语的肯定句译成英语的否定句

我们常说的正话反说指的就是这种情况，例如：

（1）你做实验要特别小心。

You cannot be too careful in doing experiments.

（2）这仅仅是个开端。

It is nothing more than a beginning,

（3）这些细菌要在温度达到一百摄氏度才会死亡。

These bacteria will not die until the temperature reaches 100℃.

"直到……才"句型译成英文一般用not...until句型。

4. 双重否定

双重否定表示肯定的意思，而且语气更为强烈，汉译英时一般都保留双重否定的形式。例如：

动物没有水不能生存，植物没有水也不能生长。

Animals cannot live without water; neither can plants grow.

Exercising Your Skills

Translate the following sentences into English using the skills you've just learned.

1. 未经允许，不得入内。

____

2. 在他还没来之前，我们就把所有的工作都干了。

____

3. 他绝不会做这样的事。

____

4. 一切正常。

____

5. 外面还在下雨，我们还不如在这儿过一晚。

____

Unit Project

The Silk Road has connected civilizations and has brought peoples and cultures into contact with each other from across the world for thousands of years, permitting not only an exchange of goods but an interaction of ideas and cultures that has shaped our world today. In the light of this enduring legacy, the UNESCO Silk Road Online Platform revives and extends these historic networks in a digital space, bringing people together in an ongoing dialogue about the Silk Road in order to foster a mutual understanding of the diverse and often inter-related cultures that have sprung up around them.

Get involved in the UNESCO Silk Road Online initiative:

If you are living in or have visited one of the regions along the historic Silk Road, you are invited to share your photos and stories with your classmates.

If you are going to visit the Silk Road，please share your travel plan and explain the reasons why you choose this route.

Unit 8 Keys

Unit 9 Heritage and Inheritance

Unit Guide: Traditional arts and crafts are interwoven in our culture and traditions, often shaping and influencing our communities in more ways than we can imagine. Unfortunately, many of these cultural heritages are slowly dying as interest continues to diminish. Yet there is still hope as a special group of people have recognized that these precious arts must be preserved and they have the inspiration and foresight to revive and reinvent these special crafts. In **Passage 1** of this unit, we will focus on preservation of the ancient craftsmanship of making Nanjing Yunjin brocade, the most extravagant silk fabric and one of the world's intangible heritages from China. In **Passage 2**, we will look at how Fan Yanyan, a silk-scarf designer in China, is drawing "authentic" inspirations from artistic treasures along the Silk Road to create her own works of art through a "combination of cultures and silk."

Lead-in

视 频

I Hit drama portrays stylish cultural heritage.

Watch the video and fill in the blanks with appropriate words.

1. Where did the Story of Yanxi Palace happen?
   The television series, set in 18th century Beijing during the Qing Dynasty, portrays the world of concubines in ________________________________.

2. What is fascinating in this television series?
   The viewers were not only fascinated by the stories but also drawn to ______________.

3. How were the costumes in the series made?
   Traditional craftsmen ________________ the fabulous costumes.

4. What artefacts have been displayed in the television series?
   The television series has put together so many ________________ artefacts.

5. Why did the hand fans shown in the television series go out of stock on online shopping platforms?

   The fans went out of stock on online shopping platforms because people ______________.

6. Where do the designs of hair accessories come from?

   The exquisite hair accessories the concubines wear were based on designs preserved in ___________________ in Beijing.

II Thirty-nine Chinese elements officially recognized by UNESCO (July, 2018).

Watch the video and fill in the blanks with the exact words you hear.

With the 24 Solar Terms, 39 Chinese cultural items have now been inscribed on UNESCO's Representative List of Intangible Cultural Heritage. Here's a full rundown of what's been included so far.

视频

31 of these Chinese elements were inscribed on The Representative List, including:

China's Twenty-Four Solar Terms—a centuries old calendar system based on 1. ______________ ____________;

Chinese Zhusuan—a 2.________________________________ using the Chinese abacus;

Chinese 3. __________________________;

The use of acupuncture and moxibustion as part of 4.______________________________ ___________;

Peking opera;

Chinese 5.________________________________ ;

Chinese engraved block printing;

Chinese 6.______________;

Chinese 7.______________________;

8.______________________________________ using timber-framed structures;

Nanjing Yunjin brocade craftsmanship;

China's Dragon Boat festival;

The Farmers' dance performed by 9.__________________ ethnic group;

...

Seven were inscribed on The List of Intangible Heritage in Need of Urgent Safeguarding. They are:

Hezhen Yimakan storytelling;

...

Chinese traditional wooden arch bridges;
Li textile techniques, which include 10.____________________________________.
And one cultural item, Chinese Fujian puppetry, was inscribed on UNESCO's List of Intangible Heritage on Register of Good Safeguarding Practices.

Ⅲ Conservation of embroidery.

Watch the video about the restoration of a piece of Dunhuang silk embroidery in British Museum and discuss the following questions.

视 频

1. What is cultural heritage? Please try making a list of cultural heritage in the world.
2. How can new technologies support the protection and promotion of cultural heritage?

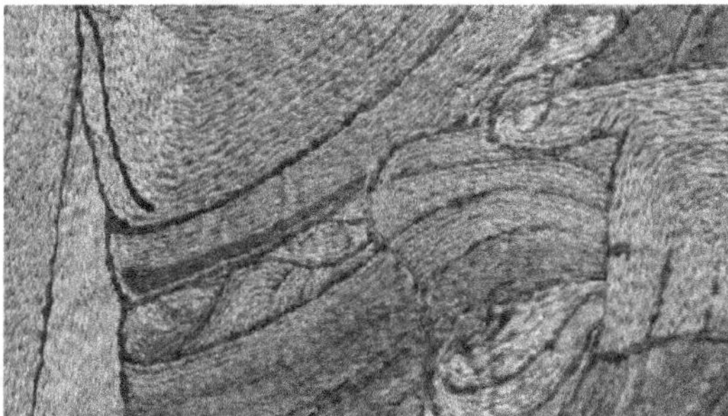

Passage 1

Passing Down the Ancient Craftsmanship of Making "Yunjin"

Li Ping

Boasting a history of more than 1,600 years, the Nanjing Yunjin brocade[1] is considered the most **extravagant** silk fabric worn by emperors in ancient times. Its Chinese name Yunjin **literally** refers to the cloud-like **splendour** of the fabrics.

A Qing Dynasty black satin with red silk embroidery in patterns of dragon, phoenix and flower.

The Nanjing Yunjin Museum[2], located in the eastern Chinese city of Nanjing, is dedicated to preserving the age-old weaving technique and creating a platform to promote textile exchanges in and outside China. The museum **showcases** fabrics, royal robes from Ming and Qing dynasties, replicas of the ancient fabrics, as well as techniques of the time-honoured weaving art.

The craftsmanship of Yunjin brocade was included on the UNESCO's **Intangible** Cultural Heritage List in 2009.

For more than 700 years, weavers in Nanjing, capital city of east China's Jiangsu Province, have been making brocades in a unique way.

Model of ancient workers making Yunjin brocade on a loom.

On the **reverse** side of brocades they apply golden threads and **flosses** of varied colours. The wooden loom they work on is 5.6 metres long, 1.4 metres wide and 4 metres high.

Visitors to the Millennium Art Museum of China Millennium Monument now have access to a traditional loom, called a **Dahualou**. They can see how it helps workers to create the famed Yunjin brocade, or "Cloud Brocade."

The loom is **on display** at an exhibition of science and technologies which runs until October 25 at the museum. It is made from 1,924 pieces of wood and bamboo, said Zhou Shuangxi, a researcher with the Nanjing-based Yunjin Brocade Research Institute.

The loom was used to make "dragon robes[3]" for emperors and empresses in Yuan, Ming

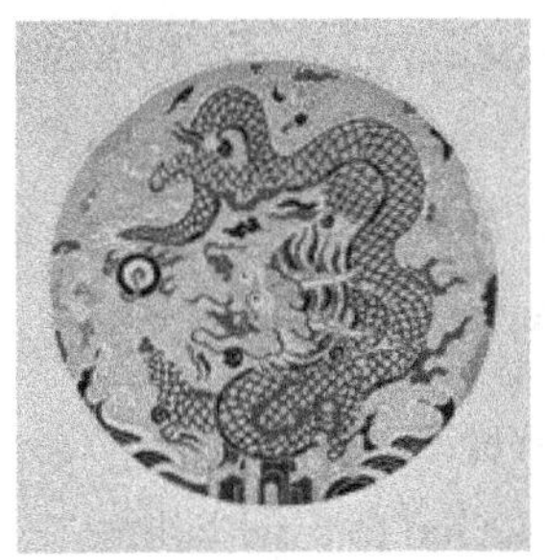

Details of a dragon pattern on a Yunjin brocade.

and Qing dynasties and also the **uniforms** of their government officials.

"The making of Yunjin involves the most complicated and difficult skills among ancient Chinese silk fabrics," said Zhou.

"The skills of weaving the Yunjin brocade is developed from two other kinds of brocades—the **Shujin** brocade[4] of Chengdu, capital of Southwest China's Sichuan Province, and **Songjin** brocade[5] of Suzhou in Jiangsu Province," he added.

The weaving work turns out to be a co-operation between two weavers. One sits on the top of the giant loom and lifts one end of a thread, and the other weaver, seated in front of the loom, weaves the other end of the thread into the brocade.

Figurines show ancient people making Yunjin brocade.

The weavers decide which thread is to be lifted up and woven into the brocade according to the pattern designed beforehand. Their co-operation results in an exquisite brocade known for its rich colours and **refined** designs.

It usually takes more than two years to accomplish one large piece. And the weaving job cannot be replaced by machines.

"Scientists have not figured out a computer program **sophisticated** enough to weave such a brocade," said Zhou.

"The secret of **manual** work defeating the high-tech world lies in ancient know-how that has been handed down by generations, " he added.

The knowledge is about such aspects as colour matching and thread lifting.

There are even instructions about knotting the threads. They tell how to position one's fingers when making a knot.

"When I became an **apprentice** in 1973, my teacher told me to start by memorising the skills," said Zhou.

Zhou's teacher had then woven brocades for half a century. "Most of the folk artists like my teacher are from poor families and they are often **illiterate**," said Zhou.

Figurines show ancient people making Yunjin brocade.

"There was a saying that if you behaved badly to your parents, or if you had no talent, you had to be a hard-working weaver in workshops. It was regarded as a punishment to be a brocade weaver," he added.

The "punishment" proved its **severity** through days of high concentration. A slight mistake led to total failure.

And if the mistake happened with a royal robe, **execution** was probable. Other punishments included being forced in a hole dug into a piece of wood attached to the loom.

"The weaving process demands a **moist** atmosphere, which means workers had to bear the **sweltering** summers and winter chills without air-conditioning," said Zhou.

Back pain and **rheumatism** were common among weavers at that time.

"My master took in six apprentices then. Only one stayed, and it was me."

Now Zhou's apprentices can produce gorgeous brocades in a comfortable environment.

Zhou is most concerned that quite a few weaving skills have been lost.

"Some were forgotten, but many of the skills were deliberately **omitted** by skilled elderly weavers in their teachings, as they were afraid of being challenged by their apprentices," said Zhou.

"Things have changed. I teach my students all I have learned. And we even have girl weavers involved, which was impossible in ancient times. Women were rejected in the area," he added.

Zhou is also worried that hand-made Yunjin brocade is being **swamped** by mass-produced replicas.

Yunjin attire and souvenirs on display on May 21, 2018.

"Like all hand-made fabrics, Nanjing Yunjin would definitely cost a lot more. A brocade of 7 meters long and 78 centimetres wide is worth at least 21,000 yuan (US$2,471)," said Zhou.

The reason for such a startling price is that two skilful weavers can only complete a small part of the fabric that is about 20 to 25 centimetres long per day on a loom.

Replicas made with machines are often **coarse** and have been primary competitors of the luxurious fabric made with traditional skills, said the researcher.

Zhou has taken with him Yunjin brocades and the "Dahualou" loom around the world.

On his trip to **Oslo**, **Norway** in 2002, Zhou was shown a piece of Chinese brocade by a **Norwegian** man.

"He treasured the brocade as a family **heirloom**. Blue dragons featured on the fabric, yet the craftsmanship didn't look great."

To his surprise, Zhou found it had been kept abroad for more than 200 years and originated from Nanjing.

"It's a pity that genuine Yunjin brocades are rarely available outside Nanjing and Jiangsu Province," said Zhou. "I wish to establish a museum for Yunjin brocades in Beijing," said Zhou.

In the museum's operation room stands 12 traditional looms for production and demonstration use. Two craftspeople operate the upper and lower parts of each loom to produce textiles incorporating fine materials such as silk, gold and peacock feather yarn. The method **comprises** more than 100 **procedures**, including manufacturing looms, drafting patterns and the many stages of weaving itself.

Different colours and patterns of Yunjin brocade on display on May 21, 2018.

The technique was once used to produce royal garments such as the dragon robe and crown costume. Today, it's widely used to make high-end attire and souvenirs.

Words

boast [bəʊst] *vt.* 以有……而自豪

extravagant [ɪk'strævəgənt] *adj.* 奢侈的

literally ['lɪtərəli] *adv.* 照字面地

splendour ['splendə] *n.* 华丽

showcase ['ʃəʊkeɪs] *vt.* 在玻璃橱窗陈列

intangible [ɪn'tændʒəbl] *adj.* 无形的；触摸不到的

reverse [rɪ'vɜːs] *adj.* 反面的

floss [flɒs] *n.* 丝绵；丝线

uniform [ˈjuːnɪfɔːm] *n.* 制服

refined [rɪˈfaɪnd] *adj.* 精制的

sophisticated [səˈfɪstɪˌkeɪtɪd] *adj.* 精密的；复杂巧妙的

manual [ˈmænjuəl] *adj.* 手工的

apprentice [əˈprentɪs] *n.* 学徒

illiterate [ɪˈlɪtərət] *n.* 不识字的人，文盲

severity [sɪˈverəti] *n.* 严重

execution [ˌeksɪˈkjuːʃn] *n.* 处决

moist [mɔɪst] *adj.* 湿润的

swelter [ˈsweltə(r)] *vi.* 热得难受

rheumatism [ˈruːmətɪzəm] *n.* 风湿病

omit [əʊˈmɪt] *vt.* 省略

swamp [swɒmp] *vt.* 使陷入困难；淹没

coarse [kɔːs] *adj.* 粗织的

heirloom [ˈeəluːm] *n.* 传家宝

comprise [kəmˈpraɪz] *vt.* 包含

procedure [prəˈsiːdʒə(r)] *n.* 程序；工序

Useful Expression

on display 陈列；展出

Proper Names

Dahualou 大花楼织机（云锦的织造工具，分为上下两层，需要两名织工同时操作才能织造）

Shujin 蜀锦，又称蜀江锦，中国传统丝织品，指四川省成都市所出产的彩锦

Songjin 宋锦，全称宋式锦，中国传统丝织品，发源地为中国苏州，故又称“苏州宋锦”

Oslo （地名）奥斯陆（挪威首都）

Norway 挪威

Norwegian 挪威人；挪威语

Notes

1. Nanjing Yunjin brocade is a kind of brocade made in Nanjing, Jiangsu, China. "Yunjin" is a general term for the traditional jacquard silk fabric. It can be traced back to the officially-run weaving factory "Jin Bureau" (锦署), which produced the brocade only for the royal family. It was not until the late Qing Dynasty that the development of Nanjing Yunjin brocade entered its period of full bloom; then the products were sold to the common citizens.
南京云锦是中国传统的丝制工艺品，有“寸锦寸金”之称，其历史可追溯至417年（东晋义熙十三年）在国都建康（今南京）设立专门管理织锦的官署——锦署。如今只有云锦还保持着传统的提花木机织造，这种靠人记忆编织的传统手工织造技艺仍无法用现代机器来替代。云锦用料考究，织造精细，在继承历代织锦的优秀传统基础上发展而来，又融会了其他各种丝织工艺的宝贵经验，达到了丝织工艺的巅峰状态，代表了中国丝织工艺的最高成就，浓缩了中国丝织技艺的精华，是中国丝绸文化的璀璨结晶。

2. Nanjing Yunjin Museum, located in the eastern China city Nanjing, is the first brocade weaving art museum in China. It includes "Yun Brocade Weaving Manufacture," "Exquisite Yun Brocade Collections," "Ethnic Brocade Weaving Performance," and "Finished Product of Yun Brocade Displaying."
坐落在中国东部南京市的南京云锦博物馆是中国唯一的云锦专业博物馆、国家三级博物馆，主要展示以南京云锦为代表的中国民族织锦艺术，是“新金陵四十八景”之一。云锦博物馆展示着云锦织造工艺、明清云锦精品实物、中国古代丝织文物复制品以及中国少数民族织锦等，1500多年手工织造历史的南京云锦，以特殊的浮雕、镶嵌技艺，表达出特殊的审美境界和文化艺术魅力，反映出中华民族特有的文化内涵。

3. Dragon robes were the everyday dress of the emperors or kings of China (since the Tang Dynasty), Korea (Goryeo and Joseon dynasties), Vietnam (Nguyễn Dynasty) and the Ryukyu Kingdom. Dragons first appeared on robes in the Tang Dynasty, and the Mongols were the first to codify their use as emblems on court robes. As a result, the subsequent Ming emperors shunned them on formal occasions. Though only royals could wear dragons, honoured officials could be granted the privilege of wearing "python robes," which resembled dragons, only with four claws instead of five.
龙袍，即绣有龙形图纹的袍服，其特点是盘领、右衽、黄色。领前后正龙各条，膝部左、右、前、后和交襟处行龙各一条，袖端正龙各一条。龙袍并不是专供皇帝穿着，郡王及以上都可以穿，只是不能用黄色，其他官员只有得到皇帝亲赐才能穿着，但必须“挑去一爪”，以示区别。在明朝，经改制后的龙袍，称为蟒袍，成为明朝职官常服。

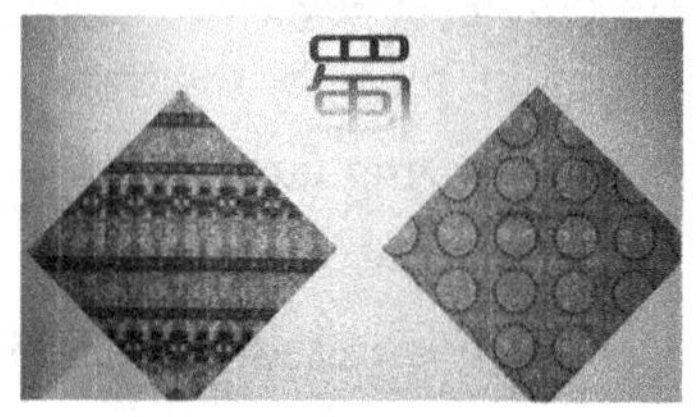

4. Shujin brocade is named after its place of origin. It dates back to the Spring and Autumn and Warrior States Period. It flourished during the Han and Tang dynasties. Shujin brocade is a multi-coloured woven fabric with Han Chinese characteristic and local influences. It is one of the four famous brocades of China.
蜀锦，指蜀地（四川成都地区）生产的丝织提花织锦。有两千年的历史。蜀锦多用染色的熟丝线织成，用经线起花，运用彩条起彩或彩条添花，用几何图案组织和纹饰相结合的方法织成。是一种具有汉民族特色和地方风格的多彩织锦。四川的蜀锦与南京的云锦、苏州的宋锦、广西的壮锦一起，并称为中国的四大名锦。

5. Named after the Song Dynasty, the period during which it was invented, the Songjin brocade is China's most luxurious and complicated form of brocade silk. The fabric is characterized by its vivid hues, exquisite patterns and luscious, soft texture. It is considered to be among the trinity of revered brocade forms that represent the epitome of silk weaving, alongside Shu Brocade from Sichuan Province and Yun Brocade from Nanjing, Jiangsu Province. But unlike the other two forms, which were typically used as fabrics for garments, the Song brocade was also used as canvases for paintings and calligraphy works. So, this particular fabric came to be associated with high society and the intelligentsia.
宋锦，因其主要产地在苏州，故又称“苏州宋锦”。宋锦色泽华丽，图案精致，质地坚柔，被赋予中国“锦绣之冠”，它与南京云锦、四川蜀锦一起，被誉为我国的三大名锦。宋锦是中国传统的丝制工艺品之一。开始于宋代末年（约公元11世纪），产品分大锦（又分重锦和细锦）、匣锦和小锦。重锦质地厚重，产品主要用于宫殿、堂室内的陈设。细锦是宋锦中最具代表性的一类，厚薄适中，广泛用于服饰、装裱。

Reading Comprehension

Answer the following questions according to Passage 1.

1. What is Yunjin brocade, according to the passage?

2. What is the purpose of establishing the Nanjing Yunjin Museum?

3. Does the Nanjing Yunjin Museum exhibit only original work of ancient Yunjin brocade?

4. How do weavers produce Yunjin brocade?

5. Can Yunjin brocade be produced massively by machines? Why or why not?

6. Who were brocade weavers in ancient time and why?

7. How hard was the brocade weaving work?

8. Why have some skills been lost?

Summary

The following is a summary of Passage 1. Fill in the blanks with appropriate words.

As the most 1. ______________ silk fabric, the Nanjing Yunjin brocade 2.____________ a long history. The weaving work involves a 3. __________________ between two weavers. They work on a giant 4.__________ loom, 5.____________ golden threads and colourful 6.__________ to the 7.___________ side of brocades according to the patterns designed 8.___________. No computer program is 9.___________ enough to weave Yunjin brocades. It usually takes more than two years to 10.___________ one large piece 11.______________.

Word Building

I The suffix *ship* is used to form nouns, meaning "quality, art or skill of; condition, rank or group of." Give the Chinese meanings to the following words containing *ship*.

kinship	____________
friendship	____________
hardship	____________
horsemanship	____________

salesmanship	____________
statesmanship	____________
relationship	____________
ambassadorship	____________
sportsmanship	____________
ownership	____________

II The prefix *co* can be used together with other words to form new words, meaning "together with, in common with or mutually." Match the following words with their parts of speech and Chinese meanings.

	cooperate	（未婚）同居
	cofounder	共谋者
	coproduction	联袂主演
n.	coexistent	合作
	cohabit	同盟；联合政府
adj.	coconspirator	飞机副驾驶员
	costar	合作生产
v.	coeducation	共同创始人
	copilot	同时代的；和平共处的
	coalition	男女同校制

Words and Expressions in Use

I Fill in the blanks with the words given below. Each word should be used twice. Change the form where necessary.

weave	thread	loom	knot	fabric	floss

1. A long time ago, when __________ was in short supply, there was not enough ready-to-wear clothing in the shops to meet public demand so people had to make their own clothing.

2. The plain porridge meal is often considered as food therapy when you have an upset stomach and it is usually served with pork ________, pickled cucumbers, and salted duck eggs.

3. The survey showed that 48.4 percent of the respondents deemed love as the foundation for tying the __________, and 40.7 percent said they wanted to enter into marriage because they love kids.

4. The fabric of Chongming cultural heritage is homespun where the three major processes are hand-spinning cotton __________, dyeing yarn and weaving cloth.

5. Chinese sci-fi writers are encouraged to __________ more Chinese elements into the genre born in the West and increase the global readership of Chinese sci-fi.

6. China underwent dramatic changes in its social __________ in the early 1990s as the country opened wider to the outside world.

7. On the "National Treasure" program, when the loom was presented on stage, the host Zhang Guoli was very excited to try it himself. He completed one __________ without any difficulty after a simple instruction by the museum staffer aside.

8. Shared bike company Mobike will replace its broken bicycles with new ones as new rules from the traffic authority __________.

9. With magic silk yarn, the Weaving Maid wove layers of magnificent clouds on her __________. The clouds that she wove were called "clothes of heaven." The clouds automatically change color according to different time and season.

10. His forehead __________ in a frown when he saw the pets in his community had caused annoyance and even injuries to residents.

11. Maria __________ her teeth every day as long as she can remember because a trip to the dentist is somcthing shc drcads most.

12. Some balls of various sizes fixed in midair. I then saw many guests in western suits bow their heads and __________ their way among the balls in a flutter of excitement.

II Render the following Chinese expressions into English. The first letter of each word is given.

1. 龙袍 d__________ r__________

2. 金线 g__________ t__________

3. 配色 c__________ m__________

4. 传家宝 f__________ h__________

5. 高端服装 h__________ a__________

6. 惊人的价格 s__________ p__________

7. 精美的锦缎 e__________ b__________

8. 大量生产的复制品 m__________ r__________

9. 历史悠久的编制艺术 t__________ w__________ a__________

10. 非物质文化遗产名录 I__________ C__________ H__________ L__________

Translation

Translate the following paragraph into English.

在古代丝织物中，“锦”是代表最高技术水平的织物。南京云锦浓缩了中国丝织技艺的精华（quintessence），是中国古代三大名锦之一。南京云锦的生产现主要分布在南京市的秦淮、建邺、白下、玄武、栖霞五区。一千五百多年前，史籍上就有关于南京丝织品的文字记载，但无实物流传。东晋末年，南京有了专门生产织锦的机构——斗场锦署（Douchang Brocade Bureau）。北宋南迁后，南京成为中国的丝织中心。南京云锦织金饰约始于元代，而彩色妆花（zhuanghua brocade）织金饰则盛于明清两代。

Passage 2

Silk's Enduring Legacy

Fan Yanyan is so **intrigued** by the Silk Road that she established her silk-scarf design studio in a Silk Road museum in Xi'an[1], Shaanxi province.

The museum's location in Northwest China was the largest market in the capital of the Tang Dynasty. It's celebrated as the traditional starting point of the Silk Road.

The fine-arts graduate believes many artistic treasures scattered along the ancient international trade route **harbour** "**authentic**" inspirations from humanity, religion and nature, rather than modern **commercialism**.

Unlike her classmates, Fan did not look for jobs after graduating from college in Xi'an in 2000. She went to Dunhuang, Gansu Province, to create **facsimiles** of **frescoes** in the Mogao Grottoes[2] for two years.

The Mogao Grottoes have more than 45,000 square meters of frescoes and 2,415 painted clay sculptures. It's the world's largest collection of historical Buddhist art, created over the period from the third to thirteenth centuries, a peak period of international trade and cultural exchanges between Central and South Asia and China.

"There were not many tourists back then. Dunhuang was a quiet and bleak town," the 38-year-old recalls.

"I felt as if I was talking with the ancient painters while making facsimiles of their works in the caves."

The two years of "lonely cultivation" deepened her understanding of art and history, and gave her inspiration, even though she was still uncertain of her future career.

Dunhuang's tourism **boom** after 2002 made her feel uncomfortable. So, she left for Beijing to work as a designer for a scarf import-export company.

"I appreciate that the job gave me a chance to paint what I learned in Dunhuang on scarves," she says.

She quit in 2008, eager for a less commercial opportunity and outlet for her talent. Before that, she got married and became a mother.

"Although I was in Beijing, my soul had never left those caves in Dunhuang," she says.

She spent a whole year at home in Beijing, resting and **contemplating** her future.

"My inspiration and passion exploded the next spring, and I knew what to do," Fan says.

Fan started her studio to paint on silk—the "best and irreplaceable material"—**crystallising** ancient China's wisdom and beauty. She painted her memories of Dunhuang on silk for 40 days straight.

"It felt like a **catharsis**, just **letting loose** the depression and inspiration that had accumulated in my mind," Fan says.

Some of the paintings created in those 40 days are exhibited in her studio in Xi'an. She believes those, which she has titled Dream Dunhuang, are her best works.

Fan and her family returned to Xi'an in 2009, which she thinks is more suitable for her creation because of the city's historical connection with the Silk Road and the Tang Dynasty—a peak in China's ancient art and literature.

She found new inspiration in the old bronze ware, the Terracotta Warriors[3], stone sculptures and many other places of historic interest in the city, which was the capital for 13 dynasties over more than 2,000 years.

Her silk painting **Twelve Chinese Zodiac Signs**—a combination of Xi'an's folk art and history—was displayed at the Shanghai World Expo 2010. There, it caught the attention of Thomas Kong, chairman of the U.S.-China Cultural Exchange and Development Association, a nongovernmental organization in Flushing, New York.

Kong flew to Xi'an to meet Fan, bought a dozen scarves she'd designed and took them to the United States.

Kong showed the scarves in an activity sponsored by the U.S. Congress. The chairwoman of a women's association asked Kong to invite Fan to design a scarf for the 16th Global Women in Leadership Economic Forum in Dubai in November 2014, as a potential official gift of the event.

Fan finished the design in one month: Women of four races surround the goddess Artemis[4], with a bright blue background that Fan favours in her designs.

Fan says the bright blue colour is a symbol of the Silk Road: Persian **merchants** brought to China many valuable **ores**, such as **lapis lazuli**[5], **calaite** and malachite, which were **ground** to make **pigments** to paint the frescoes and sculptures.

Their brilliant hues do not fade and **pose a stark contrast with** the surrounding earth-tone hue of the Gobi Desert.

"The symbolic blue means different things in the context of religion, arts and history," Fan says.

Commenting on Fan's dedication, Wu Mancong, chairwoman of the Xi'an Women Entrepreneurs Association, says: "Fan's commitment to art and her **persevering** workmanship let the ancient arts from Dunhuang enter modern people's lives.

"I hope more female entrepreneurs can make good use of their unique talents to make people's lives more artistic."

Zou Yingzi, an expert of Suzhou embroidery from Jiangsu province, says "Fan's designs integrate traditional elements and modern styles. Her dedication and wisdom make that integration look natural."

Fan recently founded a Silk Road Experience Centre in her studio, exhibiting artworks from the countries and regions along the ancient corridor.

"I hope I can draw all of these foreign cultural elements in a proper manner on silk. The combination of cultures and silk can help more people understand the **essence** of the ancient international trade route," she says.

Words

intrigue [ɪn'tri:g] *vt.* 激起……的兴趣；引发……的好奇心
harbour ['hɑ:bə(r)] *vt.* 怀有；包含
authentic [ɔ:'θentɪk] *adj.* 真正的
commercialism [kə'mɜ:ʃəlɪzəm] *n.* 商业主义
facsimile [fæk'sɪməli] *n.* 摹本；复制本
fresco ['freskəʊ] *n.* 壁画
boom [bu:m] *n.* 繁荣
contemplate ['kɒntəmpleɪt] *vt.* 考虑；思忖
crystallize ['krɪstəlaɪz] *vt.* 明确；使具体化
catharsis [kə'θɑ:sɪs] *n.* 宣泄；净化（如通过戏剧或其他艺术活动）
merchant ['mɜ:tʃənt] *n.* 商人
ore [ɔ:(r)] *n.* 矿石
calaite [kə'laɪt] *n.* 绿松石
grind [graɪnd] *vt.* 碾；磨
pigment ['pɪgmənt] *n.* 颜料
stark [stɑ:k] *adj.* 鲜明的
persevering [ˌpɜ:sə'vɪərɪŋ] *adj.* 坚韧不拔的
essence ['esns] *n.* 本质；精髓

Useful Expressions

let loose 释放，放任
pose a contrast with 与……形成对比

Proper Names

Twelve Chinese Zodiac Signs 中国十二生肖
lapis lazuli 青金石

Notes

1. Xi'an is the capital of Shaanxi Province, China, a sub-provincial city on the Guanzhong Plain in northwestern China. It is one of the oldest cities in China, and the oldest of the Four Great Ancient Capitals, having held the position under several of the most important dynasties in Chinese history, including Western Zhou, Qin, Western Han, Sui, and Tang. Xi'an is the starting point of the Silk Road and home to the Terracotta Army of Emperor Qin Shi Huang.
西安，古称长安、镐京，是陕西省省会、副省级市、关中平原城市群核心城市、丝绸之路起点城市、“一带一路”核心区、中国西部地区重要的中心城市，国家重要的科研、教育、工业基地，联合国科教文组织于1981年确定的“世界历史名城”。地处关中平原中部，北濒渭河，南依秦岭，八水润长安。

2. Mogao Grottoes, also known as the Thousand Buddha Grottoes or Caves of the Thousand Buddhas, form a system of 492 temples 25 km southeast of the centre of Dunhuang, an oasis located at a religious and cultural crossroads on the Silk Road, in Gansu Province, China. The caves contain some of the finest examples of Buddhist art spanning a period of 1,000 years. The first caves were dug out in 366 as places of Buddhist meditation and worship. The Mogao Grottoes are the best known of the Chinese Buddhist grottoes and, along with Longmen Grottoes and Yungang Grottoes, are one of the three famous ancient Buddhist sculptural sites of China.
莫高窟，俗称千佛洞，坐落在河西走廊西端的敦煌。它始建于十六国的前秦时期，历经十六国、北朝、隋、唐、五代、西夏、元等历代的兴建，石窟存有500多个洞窟，保存有绘画、彩塑492个，是世界上现存规模最大、内容最丰富的佛教艺术地。

3. The Terracotta Warriors are terracotta sculptures depicting the armies of Qin Shi Huang, the first Emperor of China. It is a form of funerary art buried with the emperor in 210 BC—209 BC with the purpose of protecting the emperor in his afterlife. The figures vary in height according to their roles, with the tallest being the generals. The figures include warriors, chariots and horses. Estimates from 2007 were that the three pits containing the Terracotta Army of more than 8,000 soldiers, 130 chariots with 520 horses, and 150 cavalry horses, the majority of which remained buried in the pits near Qin Shi Huang's mausoleum. Other terracotta non-military figures were found in other pits, including officials, acrobats, strongmen, and musicians.

兵马俑是中国第一位皇帝秦始皇的兵马俑雕塑。它是公元前210年至公元前209年与皇帝合葬的一种丧葬艺术形式，目的是保护皇帝的来世。这些人物的身高因角色而异，最高的是将军。这些人物包括战士、战车和马。据2007年的估计，这3个坑里有兵马俑8000多人，战车130辆，马匹520匹，骑兵150匹，其中大部分仍埋在秦始皇陵附近的坑里。其他兵马俑非军事人物被发现在其他坑，包括官员、杂技演员、大力士和音乐家。

4. Artemis, in the ancient Greek religion and myth, is the goddess of the hunt, the wilderness, wild animals, the Moon, and chastity. Artemis is the daughter of Zeus and Leto, and the twin sister of Apollo. She is the patron and protector of young girls. Artemis is worshipped as one of the primary goddesses of childbirth and midwifery along with Eileithyia. Much like Athena and Hestia, Artemis prefers to remain a maiden and is sworn never to marry.

阿尔忒弥斯，又名辛西亚，是古希腊神话中的狩猎女神，同时也是野兽的女主人与荒野的女领主，奥林匹斯十二主神之一。还是宙斯和勒托之女，阿波罗的孪生姐妹。阿尔忒弥斯自由独立，热爱野外生活，反对男女婚姻。喜欢和保护不嫁的处女们，她与赫斯提亚、雅典娜被视为奥林匹斯山上的3个女神。

5. Lapis lazuli or lapis for short, is a deep blue metamorphic rock used as a semi-precious stone that has been prized since antiquity for its intense colour. Lapis was highly valued by the Indus Valley Civilisation (3300 BC—1900 BC). At the end of the Middle Ages, lapis lazuli began to be exported to Europe, where it was ground into powder and made into ultramarine, the finest and most expensive of all blue pigments. It was used by some of the most important artists of the Renaissance and Baroque. Today, mines in northeast Afghanistan are still the major source of lapis lazuli. Important amounts are also produced from mines west of Lake Baikal in Russia, and in the Andes mountains in Chile. Smaller quantities are mined in Italy, Mongolia, the United States, and Canada.

青金石，其工艺名称为“青金”。在中国古代称为璆琳、金精、瑾瑜、青黛等。佛教称为吠努离或璧琉璃，是古代东西方文化交流的见证之一。资料显示，青金石是通过“丝绸之路”从阿富汗传入中国。颜色为深蓝色、紫蓝色、天蓝色、绿蓝色等。青金石还是天然蓝色颜料的主要原料。好的青金石颜色深蓝纯正，无裂纹，质地细腻，无方解石杂质。不含金星

（黄铁矿）或带有很漂亮的金星均为上品。产地有美国、阿富汗、蒙古、缅甸、智利、加拿大、巴基斯坦、印度和安哥拉等国。

Reading Comprehension

Answer the following questions according to Passage 2.

1. Where is celebrated as the traditional starting point of the Silk Road?
2. Why did Fan go to Dunhuang after her graduation?
3. How much time did it take for ancient Chinese people to create frescoes and sculptures in Mogao Grottoes?
4. How was Fan's stay in Dunhuang?
5. When did Fan leave Dunhuang and why?
6. What did she do in Beijing?
7. Why did she return to Xi'an?
8. What was another inspiration for her creation?
9. What is the symbolic colour of the Silk Road, according to Fan?
10. What does Fan hope to do?

Words and Expressions in Use

I Match the collocations in Column A with the Chinese meanings in Column B.

Column A	Column B
1.to crystallize ancient Chinese wisdom	A. 结合传统元素与现代风格
2. old bronze ware	B. 艺术珍宝
3. the starting point of the Silk Road	C. 将古代中国的的智慧清晰呈现出来
4. to integrate traditional elements and modern styles	D. 坚持不懈的工匠精神
5. artistic treasures	E. 旧铜器
6. Suzhou embroidery	F. 释放灵感
7. the persevering workmanship	G. 丝绸之路的起点
8. to let loose the inspiration	H. 苏绣

II Compare the following pairs of sentences and explain the different parts of speech and meaning of italicized words.

1. treasure
   (A) Many Spanish explorers led expeditions in search for ***treasure***, but failed.
   (B) The house was large and full of art ***treasure***.
   (C) She ***treasures*** her memories of those joyous days.
   (D) The boy is a real ***treasure*** in the team because he is so resourceful and always ready to help.

2. harbour
   (A) I've learned that when you ***harbour*** bitterness, happiness will dock elsewhere.
   (B) The house was built on the highest part of the narrow tongue of land between the ***harbour*** and the open sea.
   (C) While economic sanctions alone may not dissuade terrorists, they may cause the countries that ***harbour*** terrorists to reconsider the extent of their support.
   (D) This organization is really a ***harbour*** for women who suffer domestic violence.

3. peak
   (A) The man touched the ***peak*** of his cap and walked away.
   (B) Those snow-covered huge ***peaks*** offer viewers or climbers a fresh perspective on the world.
   (C) Off-***peak*** train times can vary across routes and train companies, but they're basically the least busy travel periods during the day.
   (D) In the middle of the Qing Dynasty, the brocade production boom ***peaked***. Along the Qinhuai River in Nanjing, the noises of people weaving could be heard day and night. Records show at that time that there were more than 30,000 looms and more than 300,000 people making a living off the trade.

4. appreciate
   (A) Your investment on Yunjin brocades should ***appreciate*** over time.
   (B) The support from the nonprofit organization is greatly ***appreciated***.
   (C) People now learn to ***appreciate*** traditional music.
   (D) The teacher avoids direct instruction and attempts to guide her students through questions and activities to discover, ***appreciate*** and verbalize the new knowledge.

5. boom
   (A) Music ***boomed*** out from the loudspeakers.

(B) On the battlefield, a soldier's hearing can be permanently damaged in an instant by the ***boom*** of an explosion.

(C) These years saw a phenomenal ***boom*** in silk products consumption.

(D) Further, high output growth among these economies has been an important factor in the global commodity price ***boom***.

6. pose

(A) At the moment, those claiming that Huawei "may" ***pose*** threats to national security have never been able to present any convincing evidence as to how the company undermines their country's national security.

(B) The gang entered the building ***posing*** as workmen.

(C) Before going into their meeting the six foreign ministers ***posed*** for photographs.

(D) A live audience will ***pose*** the questions.

Translation Skills

段落翻译（1）——段落主题要清晰

段落由多个句子组成，是更为完整的语言表达单位，段中的句子都为表达中心意义服务，句子与句子之间的意思连贯，根据内在的逻辑顺序进行安排。汉语段落可以包含一个或一个以上的主题，英语段落一般只有一个主题，通常有一个明显的主题句。例如：

（1）我们创建的合伙人机制创造性地解决了规模公司的创新力问题、领导人传承问题、未来担当力问题和文化传承问题。这几年来，我们不断研究和完善我们的制度和人才文化体系，单纯靠人或制度都不能解决问题，只有制度和人、文化完美地结合在一起，才能让公司健康持久发展。我深信，今天阿里巴巴合伙人制度和阿里巴巴所捍卫的文化，假以时日，将会越来越赢得客户、员工和股东的支持和拥护。

The partnership system we developed is a creative solution to good governance and sustainability, as it overcomes several challenges faced by companies of scale: continuous innovation, leadership succession, accountability and cultural continuity. Over the years, in iterating our management model, we have experimented with and improved on the right balance between systems and individuals. Simply relying on individuals or blindly following a system will not solve our problems. To achieve long-term sustainable growth, you need the right balance among system, people and culture. I have full confidence that our partnership system and efforts to safeguard our culture will in time win over the love and support from customers, employees and shareholders.

原文摘自马云题为“教师节快乐”的公开信，译文出自美国CNBC。原文的主题句为划线句，出现在段中，英文段落的主题句也有放在句中的情况，因此译文在语序上并没有做出调整。译文的开头做了比较大的调整，增加了双划线部分，符合英文开门见山，把重要信息方在开头的习惯。增加的部分是对“创造性地解决了”的问题的总的说明，使得读者能够迅速把握到作者想要表达的信息，也更符合英语文化的思维习惯。

（2）①中国的手工艺品、丝绸、瓷器、地毯、棉纺织品在世界上享有盛名，而且比在世界其他地方购买价格更合理，挑选余地更大。②中国各地都有自己独特的产品可买，比如说北京的景泰蓝、地毯，上海的中国服装、棉纺织品，杭州的丝绸，苏州的古玩，到西安您可以买兵马俑、唐三彩。③如果您嫌麻烦，也可以在北京、上海的友谊商店里将东西一次买齐，大城市的友谊商店一般货源都比较充足，而且一般可以代您托运。④爱好艺术的游客，还可以在中国购买字画和古玩。⑤但记住一定要有发票和正式鉴定标记（即红漆标记），才不致在海关遇到麻烦。

China has won a worldwide reputation for handicrafts, silk, porcelain, carpets and textiles, all more reasonably priced and of greater variety than elsewhere in the world.

Unique local creations are available in their own cities, such as Beijing's cloisonné and carpets, Shanghai's Chinese clothing and cotton textiles, Hangzhou's silk, Suzhou's antiques, and Xi'an's three-colored glazed pottery of the Tang Dynasty and terracotta figures.

If tourists are not able to reach these cities, most of these products can be purchased in the Friendship stores in big cities like Beijing and Shanghai. These stores will also help you transport what you have bought.

Some tourists who are keen on art may wish to buy antique objects as well as Chinese calligraphy and paintings. Be sure to get your receipts and an identification seal (a red wax seal) before you go through the Chinese customs.

（《初到中国旅游可到哪些地方》，张培基译）

原文的一个段落在译文里变成了4段，如此拆分，是因为原文段落中包含了多个主题。第一句概述了中国产品价廉物美；第二句列举了各地特产；第三句介绍了大城市的友谊商店可以买到其它城市的特产；第四、五句是对艺术爱好者的建议。鉴于英语段落一般都只包括一个主题，译文才进行了如上拆分，这样译文条理清晰，重点明确。

Exercising Your Skills

Examine the following Chinese paragraph and its English translation and explain the changes made to the English version.

对于久居在一个城市的人，放下经年累月循环往复的工作，带上家人，或约三两好友，到百里之外的地方走走看看，总是最有吸引力的选择。况且，又逢秋高气爽的季节，每年国庆节长假让人心驰神往的，就是——去外地，游山水。

不过在这样的旅游旺季出门难免会碰到诸如车票难买、住宿紧张、热门景点人山人海等问题，让你有遭遇“人灾”的烦恼。所以，选定方向和地点，提前做好各种准备，是保证出游好心情的关键。

Getting out of town is a great way for urban residents to celebrate the National Day holiday, especially those who want to spend quality time with their families. But many people will have the same idea and there will be some difficulties with your trip that can be expected such as a shortage of train tickets, crowded hotels and transit depots. So you'd better take all the possibilities into consideration before you begin your tour. Carefully developing a travel plan can ensure you have a better trip.

Unit Project

Silken Worries

Compared to its brilliant artistic and cultural history, the current prospects for Chinese silks seem very dim. This is especially true on the auction market where silk hasn't been enjoying as much popularity as other historic valuables such as jade or ceramics. While the prices of the latter two can easily break the 10 million yuan ($1.6 million) threshold, the market is still somewhat miserly when it comes to the value of historic Chinese silk products.

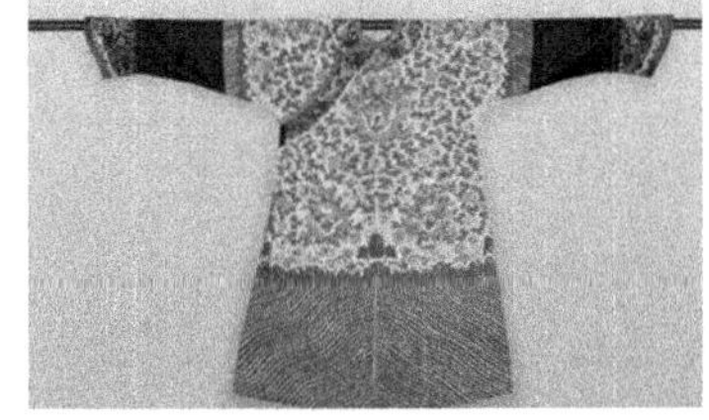

Bright Yellow Imperial Dragon Robe from the Guangxu Period (1875—1908)of the Qing Dynasty (Photo provided by Courtesy of Christie's and Council)

Unlike the rising popularity of some art work categories over the past decade, price increases for antique Chinese imperial dragon robes have been relatively stable.

"Actually, silk items don't just sit within a low price range in the auction market, they also receive little attention within academic circles as well," said Zong Fengying, an expert in silk

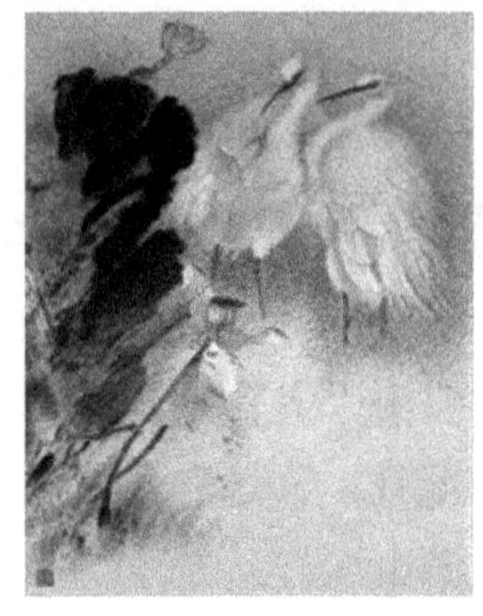

Su Embroidery The Lotus and the Crane by Sang Xingyun (Photo provided by Courtesy of Christie's and Council)

textiles at The Palace Museum in Beijing. "However, these items are extremely culturally significant."

"Silk products are presently a neglected category in the art market," said Gan Xuejun, president of Huachen Auctions in Beijing. The company has held special auctions for Su embroideries for the past two consecutive years.

"The auction results for Su embroideries have been very unsatisfying," said Gan. "It's actually very unreasonable when you compare it to their cultural and artistic value," he told the *Global Times*.

"The high cultural, academic and aesthetic value of Chinese silks will most assuredly allow them to reach market prices that accurately reflect their true value."

Work in groups to complete the following tasks:

1. What do you think of the current silken worries? Give your reasons to explain the phenomenon.
2. Work out solutions in order that the market prices of Chinese silk products truly reflect their high cultural, academic and aesthetic value.

Unit 9 Keys

Unit 10 A Tradition with a Future

Unit Guide: Silk is the most ancient and most valuable gift that China has given to the world. Though delicate looking, silk epitomizes the very spirit of the Chinese nation that looks gentle but is as strong as steel. However, silk, like many Chinese endeavors, requires intensive labor to produce. Competition from high-tech synthetics has eaten away its market share. Raw silk prices have plummeted to the point that they threaten the sustainability of this industry. In **Passage 1** of this Unit, we will look at what silk producers in China are doing to revive this ancient industry; and in **Passage 2**, we will see what collaborative efforts are being made in the major silk producing countries in Asia to promote the silk industry. Working together with the ever-green spirit of the Silk Road, "peace and cooperation, openness and inclusiveness, mutual learning and mutual benefit, " we are sure that this traditional and environmentally sustainable product will enjoy a silky smooth journey of revival.

Lead-in

I Sericulture moving from East to West.

With the economic development and aging of sericulture farmers, sericulture production in Eastern China (Yangtze River Delta and Shandong) has shrunk sharply. Addressing to this problem as well as echoing the policy of Poverty Alleviation and Rural Vitalisation, the Ministry of Commerce of China issued *The Project of Sericulture Moving from East to West* in 2006.

Step One: Please describe the moving of sericulture from East to West.

Step Two: New and innovative modes of sericulture are adopted to improve the quality of cocoons as well as the efficiency. Please translate the underlined expressions into Chinese.

1. The family-farm mode of sericulture is adopted to achieve the uniformity of seed supply, young silkworm breeding, technical services, and purchase of cocoons according to standards.
   The family-farm mode of sericulture: ________________

2. The young silkworm breeding base + farmer mode of sericulture refers to the unified breading of young silkworms to 2 or 3 moults in the base and then the silkworms are distributed for the farmers to raise under the guide of professional technicians.
   The young silkworm breeding base + farmer mode of sericulture: ________________

3. The whole-age artificial diet mood of sericulture is developed to achieve low-cost feeding formulations, mechanised feeding lines, and disease prevention and control.
   The whole-age artificial diet mood of sericulture: ________________

II Clothing ethics.

Watch the video and fill in the blanks with the right words in the brackets.

视 频

1. Ma Ke is a designer __________ (for / against) fashion, and her designing has nothing to do with fashion, but intends to present the spirit behind the clothing.

2. Traditional Chinese folk handicrafts should be recorded and then __________ (passed down / forgotten). Over time, people will gradually realize what these handicrafts really reflect—the most precious thing in life.

3. While fashion encourages people to keep buying and abandoning, handicrafts can help people find a sense of satisfaction and happiness and then consume __________ (more / less).

4. The more open our society is, the more it is the right time for us to find our own position. We advocate an ______________ (environment-friendly / user-friendly) style of clothes-making with zero-carbon emission.

5. When Chinese culture goes global, it will make the world ______________ (more / less) harmonious, peaceful, and tranquil.

III Extraordinary applications of silk.

Except that silk makes good material for clothes, accessories and textiles, it enjoys a wider range of applications in the fields of industry, medicines and health care. Study the following pictures, guess what the items are, and explain their functions. Besides, can you figure out other applications of silk?

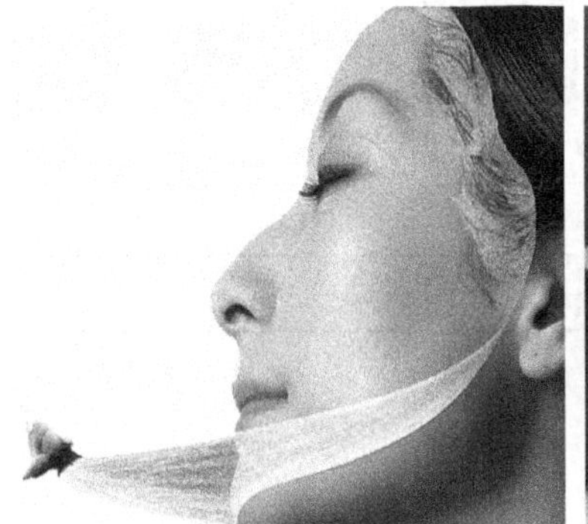

Picture 1

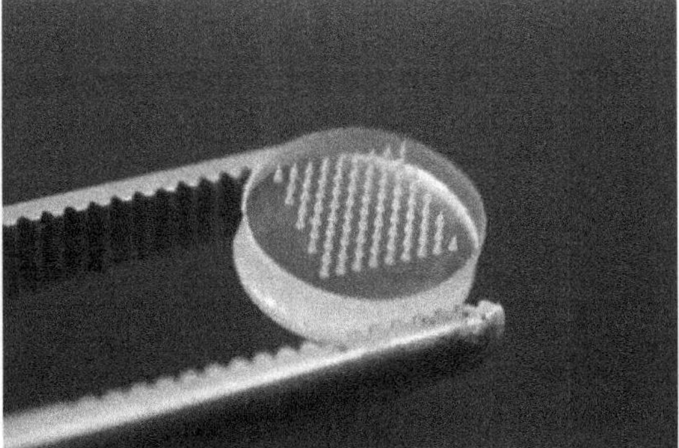

Picture 2

______________________________ ______________________________

______________________________ ______________________________

Picture 3

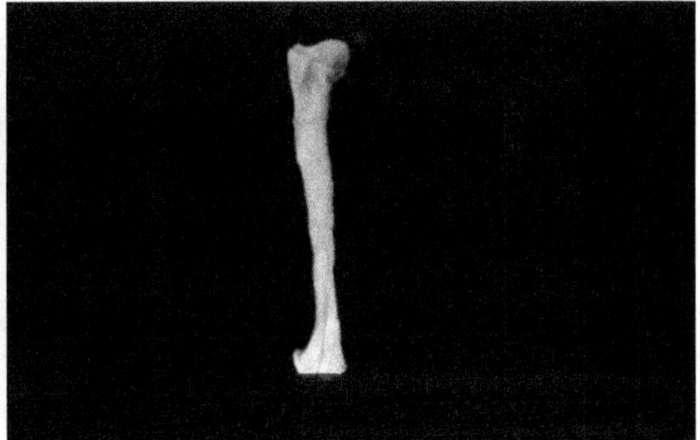

Picture 4

______________________________ ______________________________

______________________________ ______________________________

Picture 5

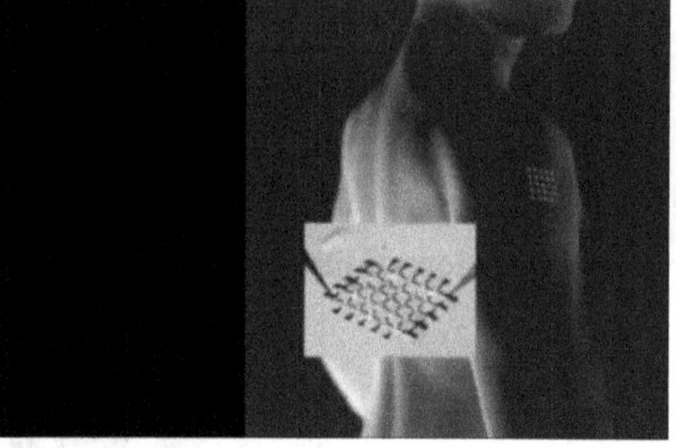

Picture 6

Passage 1

Silky Smooth Industrial Revival

Xu Junqian

The ancient silk town of Huzhou in East China's Zhejiang Province, from which exquisite fabrics once **clothed** Chinese emperors and British queens, is now one of the key industrial cities hosting thousands of factories that **churn out** a wide range of consumer goods from stuffed toys to flat-panel TV sets. Amid rapid industrialization, the art of silk making has largely been lost.

While most people in Huzhou have accepted the town's transformation, local **entrepreneur** Ling Lanfang has spent most of his time and money trying to **revive** the trade.

The 58-year-old former silk-spinning worker bought a bankrupt silk workshop in 1998 with his entire savings of 28,500 yuan ($4,341).

Since then, he has turned it into a money-making enterprise, which he named Silk Road[1], through a variety of businesses including building materials and **pharmaceuticals**.

In his office in an old-fashioned four-floor building at the city centre, Ling said that his primary interest remains silk.

"I have been **sinking** a big portion of the profits from my other businesses into silk making with the hope of reviving the lost glory of my home town," Ling said.

Indeed, silk put Huzhou on the international map when samples of Jili[2] silk, woven by the town's best craftsmen, won the gold medal at the first World Exposition in London in 1851.

"Now I just want to bring back the lost craftsmanship and re-establish a **thriving** silk industry that was our pride for many centuries," Ling said. "It's not all about money."

His efforts are beginning to produce results. Last year, representatives from some of the world's top fashion houses such as **Chanel** and **Hermes** placed test orders for his silk products.

Like many upcoming cities in China's affluent **Yangtze River Delta**, Huzhou, covering an area of 5,818 square kilometres, boasts a soaring property price that hardly matches its residents' income and a growing skyline of high-rises that follow the metropolis lying 130 kilometres northeast of it, Shanghai.

But the 2,300-year-old city also boasts a long tradition of sericulture that has fuelled the pride of many of its residents like Ling, one generation after another, and made them believe it was their ancestors that have built up today's Shanghai, the financial hub, to some degree.

According to Lu Shihu, a local historian, people in Huzhou began planting mulberry trees and raising silkworms in the Yuan Dynasty.

Its geographical position by the Taihu Lake and warm climate made it an ideal base for sericulture, producing silk that is **distinguished** by a milky-white colour.

The fabric is **resilient** and **translucent,** and was favoured by the imperial court in Beijing.

Recent archaeological findings, a few pieces of fragment silks dating from 4,700 years ago discovered in Huzhou, also proved that it is the earliest place to have carried out sericulture.

It was not until the 1840s when foreign nations were allowed to trade in Shanghai that the industry started to prosper.

Huzhou silk established itself as being synonymous with the best silk in the world.

Large quantities of silk were shipped from harbours in Shanghai to European countries, where silk was still a very precious and rare material for royals.

Wealth quickly **accumulated** among several Huzhou silk **tycoons**, who had **taken up** more than half of the market then.

Nobody **keeps count of** the combined volumes of silk exported in those large wooden containers. But in the following years, as the silkworms kept by almost every family in Huzhou spun cocoons and then silk, villas, **voluptuous** ball rooms and a 24-floor-tall skyscraper[3], the tallest then in Asia, were built in Shanghai by Huzhou businessmen.

It is estimated that the collective wealth of these silk tycoons reached more than 80 million liang of silver[4] (2.5 million in kg, equals 16 billion yuan), **exceeding** the annual revenue of the central Qing government then.

Even in the **secluded** days of China in the 1960s, the silk industry in Huzhou was **at its zenith**. Every one out of five people in the city was working in the industry, and every year the industry created a total amount of foreign currency for the country that could build another six Huzhou cities, said Zhang Jianxin, a retired director of the No. 2 Zhejiang Silk Plant, one of the biggest textile factories in China then.

But shipping silk **in bulk** abroad no longer seems to be such a profitable business in recent years.

Even though 10 percent of the silk in China, the world's largest silk producer, still come from Huzhou today, Ling, the president of the Silk Road Group, the biggest silk manufacturer in the city which also ranks third in the country with a market share of 3 percent, has found himself profitless.

In 2010, the group **grossed** more than 2 billion yuan. Yet, of its 60-million-yuan profit and tax, only 33 percent came from the silk business, which employed up to 90 percent of its 3,000 employees. The majority of the profit came from **subsidiary** businesses such as medicines and building materials.

"When I invested all of my savings, 28,500 yuan, to **buy out** the silk factory that **laid** me **off** 20 years ago, I never expected it to be so difficult," said Ling, who started his silk career right after finishing junior high school, at the age of 17.

Customs statistics show that China's silk exports exceeded $3 billion in 2010, and Zhejiang remains the largest producer, **accounting for** about one-fifth of the total exports.

Silk trade with Italy, China's third-largest silk export destination after the United States and India, enjoyed a 37.7 percent increase from 2009, the highest in terms of growth rate among all countries, according to Customs.

But now with an increase of 15 percent of revenue from silk every year, Ling seems more than optimistic about the future of his silk business.

"Hermes sells its square-metre-sized silk scarf for 2,000 yuan a piece. The price of **raw silk** exported from here is 200 yuan a kilogram, which has more than **tripled** in the past few years," Ling said.

"If we keep being the middleman squeezed between the cocoon farmers and the foreign companies, we are sure to lose. But if we sell made-up silk articles as our ancestors sold raw silk century ago, it's a whole new story. It's not like making computers or building planes, we are known for being good at it."

Ling said he has been investing 50 million to 60 million yuan every year since 2008 to upgrade his machines and buy out local factories to **complement** his business scope.

"This year, I want to set my foot abroad, seeking more opportunities with experienced design workshops in Italy or France, the centre of fashion world, and we would like to talk to anyone interested," Ling said.

"Now we have the best material, the most skilled hands, and the most **high-end** machines. All we need is a brilliant stroke to make our milky-white silk more colourful and we are a step closer to being the '**Louis Vuitton**' of the silk world."

Words

revival [rɪˈvaɪvl] *n.* 复兴，复活
clothe [kləʊð] *v.* 穿衣服
entrepreneur [ˌɒntrəprəˈnɜː(r)] *n.* 企业家
revive [rɪˈvaɪv] *vt.* 使复兴，使复活
pharmaceutical [ˌfɑːməˈsuːtɪkl] *n.* 药物
sink [sɪŋk] *vt.* 投入（资金）
thriving [θraɪvɪŋ] *adj.* 兴旺的；繁荣的
distinguished [dɪˈstɪŋgwɪʃt] *adj.* 著名的；卓越的
resilient [rɪˈzɪliənt] *adj.* 能复原的；弹回的；有弹性的
translucent [trænsˈluːsnt] *adj.* 半透明的；透亮的，有光泽的
accumulate [əˈkjuːmjəleɪt] *vi.* 堆积；积累
tycoon [taɪˈkuːn] *n.* 大亨；企业巨头
voluptuous [vəˈlʌptʃuəs] *adj.* 骄奢淫逸的；倾向感官享受的
exceed [ɪkˈsiːd] *vt.* 超过；超越；胜过
secluded [sɪˈkluːdɪd] *adj.* 与世隔绝的；偏僻的
zenith [ˈzenɪθ] *n.* 顶点，极点
gross [grəʊs] *vt.* 总共收入
subsidiary [səbˈsɪdiəri] *adj.* 附带的；附属的
triple [ˈtrɪpl] *vi.* 增至三倍
complement [ˈkɒmplɪment] *vt.* 补足；补充；补助
high-end [haɪ end] *adj.* 高端的；高档的

Useful Expressions

churn out 快速生产；大量生产
take up （在数量、比例上）占据
keep count of 计数，计算
at its zenith 处于顶点，处于鼎盛期

in bulk 大量；整批
buy out 买下……的全部股份；赎买
lay off 暂时解雇，裁员
account for （在数量、比例上）占据
raw silk 生丝

Proper Names

Chanel 香奈尔，法国品牌，创始人Coco Chanel
Hermes 爱马仕，法国品牌，创始人Thierry Hermès
Yangtze River Delta 长三角地区
Louis Vuitton 路易•威登，法国品牌，创始人Louis Vuitton

Notes

1. Here, Silk Road refers to the Silk Road Group founded by Ling Lanfang. Headquartered in Huzhou, Zhejiang, the Silk Road Group has fifteen subsidiary companies across China and is a major producer of silk making, digital weaving, and fashionable home textiles.
这里的丝绸之路指的是凌兰芳创立的丝绸之路集团公司。丝绸之路集团公司总部位于浙江湖州，在全国拥有15家子公司，主要从事制丝 、数码纺织和时尚家纺品的研发和生产。

2. Jili is a village of Huzhou, Zhejiang province. According to historical records, Jili village began to raise silkworms and reel silk in the end of the Yuan Dynasty, and in the middle period of the Ming Dynasty, silk made in Jili was chosen to be the material of the emperors' clothes.
辑里是浙江省湖州市的一个村。据史料记载，辑里村在元朝后期开始养蚕织绸，到了明朝中期，辑里丝绸成为帝王御用的衣料。

3. Here, the 24-floor-tall skyscraper refers to the Park Hotel on Nanjing Road West, Shanghai. It was the tallest building in Asia from its completion in 1934 to 1958.
这里，这幢24层高的摩天大楼指的是位于上海南京西路上的国际饭店。该大楼在1934年到1958年是亚洲的最高建筑。

4. Liang of silver is a weight unit in ancient China. 16 liang of silver=500 g.
两是中国古代的重量单位。16两=500克。

Reading Comprehension

Decide whether the following statements are true or false. Write "T" for true, and "F" for false.

1. ( ) The silk fabrics made in Huzhou used to be the clothing material for the royal families.

2. ( ) Ling Lanfang founded the Silk Road Group mainly in the hope of making money.

3. ( ) Huzhou started to enjoy an international name for silk because of Jili silk winning the gold medal at the first World Exposition.

4. ( ) Two factors contribute to the uniqueness of Huzhou silk: the geographical location and the warm climate.

5. ( ) It was the prospering silk trade that helped to build Shanghai as a flourishing city.

6. ( ) In the 1960s, the silk industry in Huzhou underwent a drop and lost its glory.

7. ( ) Recent years witness a declining silk industry because of the decreasing labour force.

8. ( ) Ling Lanfang is very optimistic about the future of the silk industry of Huzhou.

Summary

Fill in the blanks with appropriate words. The first letter of each blank is already given.

Huzhou, with its special 1. g__________ location and warm weather, is ideal for sericulture, and its 2300-year-old tradition of sericulture has 2. f__________ the pride of many local people.

Huzhou silk used to enjoy great glory. From the 1840s, Huzhou silk started to be exported to foreign countries and established itself as being 3. s__________ with the best silk in the world. Precious and rare, exquisite fabrics of Huzhou silk were used to 4. c__________ both Chinese emperors and foreign royals. The thriving silk trade brought great 5. w__________ to people who were engaged in the industry.

Yet, recent years witness a declining silk industry. According to Ling Lanfang, the president of the Silk Road Group, the biggest silk manufacturer in Huzhou, things are difficult now and his silk business is 6. p__________. Despite the difficulty, he plans to go abroad to seek more opportunities and he is sure that with the best material, the most skilled 7. c__________, the most high-end machines, and a variety of colourful milky-white Jili silk, the lost glory of the Chinese silk industry will be 8. r__________ in the near future.

Word Building

In the passage, there are some synonymous expressions. Match the expressions in Column A with their respective synonymous expressions in Column B.

Column A	Column B
invest	sink
revive	growing
take up	quantity
volume	medicines
soaring	bring back
profitable	account for
pharmaceuticals	money-making

Words and Expressions in Use

I Fill in the blanks with the words or expressions given below. Change the form where necessary. Each word or expression can be used only once.

clothe	revive	thrive	revenue	accumulate	distinguished
translucent	exquisite	zenith	lay off	in bulk	account for

1. Barong clothing is a kind of __________ and embroidered clothing worn by Filipinos at special events.

2. Since 2013, Liu Lanfang and her company's design team have been working to __________ traditional Qingyang sachets (庆阳香包) with new designs while keeping to the craft's original style.

3. During the reign of Emperor Hui Zong of Song Dynasty, the art of embroidery came to its __________ and reputed workers popped up.

4. In 2017, Gannan received a total of 11.056 million domestic and foreign tourists, with a comprehensive __________ of 5.15 billion.

5. Kesi first rose to prominence during the Ming Dynasty and __________ until the end of the following Qing Dynasty.

6. Boasting a combination of traditional techniques with modern elements, our __________ and fashionable designs of silk ribbon embroidery enjoy great popularity with the public.

7. Jaber al-Jaberi, Iraqi deputy minister of culture noted that China has __________ cultural, intellectual, economic, social positions in the world as Chinese people have been walking on the Silk Road for over 2,000 years.

8. Skiing is the most popular activity in winter, with ski resorts __________ 60 percent of the market nationwide.

9. Unearthed in the Buddhist shrine in Xinjiang is a clay head sculpture of a Sogdian gongyangren, or donor. Bearing in mind the wealth they __________ in trade, it is not surprising that some of that wealth was offered up to temples of worship.

10. It is the custom for the APEC leaders to be __________ in traditional outfits of the host country on the final day of the forum—a symbol of "Big Family."

11. Dressed simply and in light make-up, Zhang Surong looks no different from other women her age. The only difference is that, eight years after being __________ from a state-owned enterprise, she now owns four firms and possesses a personal fortune of nearly 40 million yuan.

12. Buying the traditional silk fans __________, you can negotiate a better price.

II Render the following Chinese expressions into English. The first letter of each word is given.

1. 精致的面料 e__________ f__________

2. 高端的机器 h__________ m__________

3. 排名第三 r__________ t__________

4. 世界博览会 W__________ E__________

5. 不赚钱的行业 p__________ i__________

6. 富饶的长三角 a__________ Y__________ R__________ D__________

7. 激增的地产价格 s__________ p__________ p__________

8. 积累巨大的财富 a__________ g__________ w__________

9. 复兴逝去的光辉 r__________ the l__________ g__________

10. 买入一家破产的作坊 b__________ o__________ a b__________ w__________

Translation

Translate the following paragraph into English.

辑里丝是湖州南浔的特产，被誉为丝中极品。南浔地处太湖之滨，河流纵横，水中泥土含有多种矿物质，养分丰富，提供了育桑养蚕的有利条件。辑里丝色泽润白，光洁柔韧，一根丝线能悬挂8枚铜钱而不断，因此成为丝中极品和优质中国蚕丝的代名词。据史料记载，明清两朝的帝王皇袍必须用辑里丝做。如今，南浔辑里湖丝馆还保存着两卷60年前的辑里湖丝，质地仍鲜艳如初。

Passage 2

Silk Profession and Culture of Silk in Modern Life–Future Development of Asia Silk Alliance

Dilip Barooah[1]

Products that require more human touch and input are likely to be more and more expensive over time. This would **remain applicable to** absolutely specialized, hand crafted products with human inputs of high skill. Sericulture and silk products **fall into** this category. The host plant farming to rearing of silk shall continue to demand specialized skills and human inputs besides, the technological contribution. A silk product weighing just 100 gram means an output of **nurturing** of as many as 450 **nos**[2] of silk worms over a period of about 21 days cycle; that too, up to the cocoon stage only. By the time the raw silk turns out to be a finished product, the human touch multiplies and eventually that becomes a luxury product.

Purchasing power across the globe is increasing with the rise in standard of living. Even the poorer areas of the earth, the growth has been remarkable. Perception of "luxury" is changing. Yesterday's luxury may not be today's luxury. Time needed, even for a common man to earn the purchase price of expensive luxury items is diminishing (Source: *1895* ***Montgomery Ward*** [2] *catalogue*). At the same time, environmental concerns are growing. Popularity of natural and low impact textiles such as silk, linen is increasing. Currently, traditional silk market is in the **upward trend**.

Despite the fact that silk enjoys only 0.5% share in the global fashion trade, in value terms the market size is as big as $ 1.15 Trillion. With such increasing potential, silk manufacturing countries should aim to **optimize** the resource **capitalization**.

Today's largest producers and suppliers of raw silk and silk yarn are in Asia, **with the** notable **exception of** Brazil. Sericulture and silk production continues to be labour-intensive and remain at the village level, employing both men and women **at all stages of** production. In China, this sector occupies some 20 million farmers, as well as 500,000 people in the silk processing industry. In India, sericulture is a cottage industry in 59,000 villages. As one of the most labour-intensive sectors of the Indian economy, it provides full and part-time employment to some six million people. In fact, this sector has **been identified as** a sector of the Indian economy with strong potential to create jobs.

Asia will continue to be the driving force for silk, not only in terms of production, but also in terms of consumption, due to the traditional **attachments** and love for silk across the continent. As growing together is the most sustainable recipe, we need to promote silk **collaboratively** with our sincere efforts, for which **alliance** and collaboration are essential elements.

With this **holistic** approach Asia Silk Alliance (ASA) was formed, as an initiative by The **Mekong** Institute[3] in collaboration with the **Khon Kaen** Provincial Government, Thailand, in the year 2015, to **spearhead** silk promotion through organizing International silk seminars. It aims to develop silk industry in the region. Representatives from Cambodia, China, India, Japan, Laos, Myanmar, Thailand and Vietnamhave have pledged to continue this project **on a voluntary basis**.

Some Key Notes of Asian Silk Alliance (ASA)

Vision: Building partnership for success of silk in Asia.

Mission:

Bring together and develop ASIA Silk Network among silk **stakeholders** i.e. producers, suppliers, importer, exporter and customers.

Work in partnership with each other to help develop silk communities in each region.

Vie for responsible and inclusive silk business-practices, driving product development through green and innovative value chain.

Advocate silk business solutions for Asia challenges.

Services:

Encourage all stakeholders and interest groups to join "Asia Silk Strategic Partner Programme" to connect with the leading companies across Asia who share best practices and challenges in silk business.

Promote silk products among the group member that aim to expand into the global market.

Provide the group member with a space to share, discuss, learn and collaborate. Specific focus is given to impact measurement using tools of business strategies by organizing annual meeting event such as 'Silk Sector Round Table, ' International Seminar or Asia Silk Sector Forum.

Anticipate global challenges that affect silk business sector, such as consumer trends, international economic situations, technology and innovation adaptation in order to response promptly to meet the market challenges and demand.

Each silk producing country has their respective strengths, weakness, opportunities and threats (SWOT). It is now time that we **initiate** an approach to design a **sustainable** SWOT Based Package that is market-friendly across the globe. Let the "strength" and "opportunities" of one country plays the role to **mitigate** the "weakness" and "threats" of another country.

China being the leader, the silk-world would appreciate a situation where China plays the role of the driving force to see that the entire stake holders of the world of silk, **irrespective of** political barriers, if any, can enjoy a sustainable business. This will nevertheless **bolster** China's position even stronger. With this endeavour, the Preparatory Meeting and Assembly of International Silk Enterprises Promoting Development Union was held on October 23rd 2015, led by China's Zhejiang Cathaya International Co., Ltd.[4] Silk enterprises around the world are considering collectively to see the possibilities of carrying out more **inter-operability**, strengthen the exchange of industrial information, promote industry trade and investment between the countries and ensure **sustainability** of the global silk industry. More than 70 silk enterprises and related organizations from China, Italy, France, Thailand, India, Vietnam, Singapore, Brazil, **Burma** and Japan etc. totally 13 countries and regions have already joined the International Silk Union (ISU)[5] voluntarily.

The International Silk Union is an international and non-profit social organization with the main focus towards development of silk industry. The foundation of ISU is supported by The National Chrysalis Silk[6], Chinese Silk Association[7], Zhejiang Province Economy and Information committee, the Commerce Department, Hangzhou Municipal Government.

The president of Zhejiang **Cathaya** International Co., Ltd, Mr. Li Jilin is elected as the chairman and the director of Hangzhou Silk Culture and Brand Research Centre. Mr. Fei Jianming, also an active country lead of ASA, is elected as Secretary General of the union. The director of China Silk Association, Mr. Yang Yongyuan is invited as honorary chairman of the union on Chinese side.

Challenges and opportunities ahead:

International partnerships often encounter barriers such as resource, capacity, political and cultural differences which affect the motivations, balance of benefits, and ultimately outcomes of such collaboration.

As there are no restrictions in silk and silk products in the international trade, a massive international campaign to **popularize** silk under a socially conscious brand (like fair trade, organic etc.) is the need of the time.

Conclusion:

The ever green spirit of the Silk Road—"peace and cooperation, openness and inclusiveness, mutual learning and mutual benefit" is now going to play a great role in the modern life, to promote prosperity and development of the countries along the Silk Road, with advanced connectivity. Although, it will **abide by** market rules and international **norms**, building bridges between people and their partnership through ASA and ISU will lead us to our silk journey.

Words

applicable [ə'plɪkəbl] *adj.* 适当的；可应用的
nurture ['nɜːtʃə(r)] *vt.* 培育；养育；培植
nos ['nʌmbəz] *abbr. numbers* 数
optimize ['ɒptɪmaɪz] *vt.* 使最优化，使尽可能有效
capitalization [ˌkæpɪtəlaɪ'zeɪʃn] *n.* 资本化
attachment [ə'tætʃmənt] *n.* 附属物；依恋；依附
collaboratively [kə'læbərətɪvli] *adv.* 合作地；协作地
alliance [ə'laɪəns] *n.* 结盟；同盟；联盟；联合
holistic [həʊ'lɪstɪk] *adj.* 整体的；全面的
spearhead ['spɪəhed] *vt.* 做……的先锋；带头做
voluntary ['vɒləntri] *adj.* 自愿的；主动的
stakeholder ['steɪkhəʊldə(r)] *n.* 利益相关者
vie [vaɪ] *v.* 激烈竞争；争夺
initiate [ɪ'nɪʃieɪt] *vt.* 开始，创始；发起
sustainable [sə'steɪnəbl] *adj.* 可持续的
mitigate ['mɪtɪgeɪt] *vt.* 使缓和；使减轻
irrespective [ˌɪrɪ'spektɪv] *adj.* 无关的；不考虑的；不顾的
bolster ['bəʊlstə] *vt.* 支持；支撑
inter-operability ['ɪntərˌɒpərə'bɪlɪti] *n.* 互操作性；互通性
sustainability [sə'steɪnəbɪləti] *n.* 可持续性；永续性
popularize ['pɒpjʊləraɪz] *v.* 推广；宣传；使普及；使流行
norms [nɔːmz] *n.* 规范；准则；行为标准

Useful Expressions

remain applicable to　仍适用于……
fall into　落入；分成
upward trend　上升趋势
with the exception of　除……之外
at all stages　在各个阶段；在所有阶段
be identified as　被认为是；被看作；被确认是
on a voluntary basis　按照自愿原则；出于自愿
vie for　争夺；竞争；抢
irrespective of　不考虑；不管；不受……影响
abide by　遵守；恪守；遵从；遵照

Proper Names

Montgomery Ward　蒙哥马利·沃德百货公司
Mekong　（地名）湄公河
Khon Kaen　（地名）孔敬，泰国依善地区的一个城市
Myanmar　（国家名）缅甸，即Burma
Cathaya　（公司名）凯喜雅

Notes

1. The author, Dilip Barooah, is the Managing Director at Fabric Plus Pvt. Ltd. and Rudrasagar Silk Ltd.
作者迪利普·巴罗阿是Fabric Plus私人有限公司及鲁德拉萨加尔丝绸有限公司的执行经理。

2. Founded by Aaron Montgomery Ward in 1872, Montgomery Ward Inc. is the name of two historically distinct American retail enterprises. It can refer either to the defunct mail order and department store retailer, which operated between 1872 and 2001, or to the current catalog and online retailer also known as Wards. *1895 Montgomery Ward catalogue* is an illustrated mail order catalogue published by Montgomery Ward selling over 20,000 items.

蒙哥马利·沃德公司由阿伦·蒙哥马利·沃德于1872年在芝加哥创立，这个名称下有两家公司，一家是经营邮购和百货公司业务的公司，已于2001年倒闭，现在的公司名称是沃德（Wards）主要业务是邮购和网络销售。*1895 Montgomery Ward catalogue*是蒙哥马利·沃德公司1895年出版的一本邮购目录，里面有20000多种产品。

3. Mekong Institute (MI) is an Intergovernmental Organization (IGO) which provides, implements, and facilitates integrated human resource development (HR(D) and capacity building programs and development projects related to regional cooperation and integration issues in the Greater Mekong Subregion (GMS). MI is the only GMS-based HRD and capacity building institute owned and operated by the six GMS governments; Cambodia, Lao P.D.R, Myanmar, Thailand, Vietnam and China.
湄公学院（MI）是一个政府间组织，负责提供、实施和促进大湄公河次区域的区域合作和一体化相关的人力资源开发和能力建设项目。湄公学院是该地区唯一一家由柬埔寨、老挝、缅甸、泰国、越南和中国六个国家政府拥有和运营的人力资源开发和能力建设机构。

4. Zhejiang Cathaya International Co., LTD. 浙江凯喜雅国际股份有限公司。

5. International Silk Union (ISU) is an international and specialized non-profit organization founded in October 2015 with a purpose to promote communication and cooperation of the silk industry among countries, and to promote the progress and development on international silk culture, education, scientific research, design, production, standards and testing, trade and consumption etc. The union strictly complies with the constitution, laws, regulations, national policies and also the social morality of each member's country in carrying out the activities. There have been more than 110 enterprises and institutions from over 17 countries and regions joined ISU, including China, Brazil, Italy, India, France, Japan, Switzerland, Thailand and Vietnam etc.
国际丝绸联盟是一家专业的非营利性国际机构，2015年10月成立，旨在促进各国丝绸行业的相互交流和合作，推动国际丝绸文化、教育、科研、设计、生产、标准与检测、贸易、消费等方面的进步与发展。联盟在开展活动时严格遵守各成员所在国的宪法、法律、法规和国家政策，遵守各成员所在国的社会道德风尚。目前已有来自中国、巴西、意大利、印度、法国、日本、瑞士、泰国、越南在内的17个国家和地区的110多家企业和机构加入联盟。

6. The National Chrysalis Silk 中国茧丝绸行业协会。

7. China Silk Association was founded in 1986, is a national industry association based on silk production and trade enterprises. At present, the association has a member of

more than 700 units, including national key cocoon baking, filature, silk, silk fabric, silk printing and dyeing, silk clothing, silk knitting, silk machinery enterprises and many other colleges and scientific research institutions form provinces producing silk including, Jiangsu, Shanghai, Zhejiang District, Sichuan, Shandong, Anhui, Chongqing, Guangdong, Guangxi, Liaoning etc.

中国丝绸协会是1986年成立的，是以生产企业为基础、茧丝绸贸工农相结合的全国性行业组织。协会共有会员单位700多家，其中既包括了浙江、江苏、上海、四川、山东、安徽、重庆、广东、广西、辽宁等丝绸主产区的省市丝绸公司，也包括了全国重点骨干蚕茧收烘、缫丝、绢纺、丝织、丝绸印染、丝针织、丝绸服装、丝绸机械等众多的企业以及科研院校。协会主管部门为国务院国有资产监督管理委员会。

Reading Comprehension

I Answer the following questions according to what you read from Passage 2.

1. What is this passage mainly about?

2. What kind of products does the passage say is likely to become more and more expensive over time?

3. How long does it take for a silkworm to become a cocoon?

4. What is current situation of silk in the global market?

5. Why is silk market in the upward trend?

6. Who founded Asia Silk Alliance and what is its purpose?

7. What is the role of China in the silk world?

8. What challenges are ASA facing?

9. What is suggested to combat the challenges and opportunities?

10. What kind of future about the silk industry is hoped for by ISU and ASA?

II Questions for critical thinking.

The Belt and Road Initiative was first proposed by President Xi Jinping in 2013. As a college student, what have you benefited from this initiative? What challenges and opportunities does the initiative present to you in the future when you graduate from the university?

Words and Expressions in Use

I The following are a list of acronyms important to the silk industry and silk culture. Find out the full English name of the following acronyms and their equivalent Chinese names.

Acronyms	Full English Name	Chinese Name
1. ISU	____________________	____________________
2. CSA	____________________	____________________
3. CCCT	____________________	____________________
4. NCSCO	____________________	____________________
5. CSC	____________________	____________________
6. IASSRT	____________________	____________________
7. UNESCO	____________________	____________________
8. BRI (OBOR)	____________________	____________________
9. CNSM	____________________	____________________
10. CCSE	____________________	____________________
11. NPSSTT	____________________	____________________
12. SRICAAS	____________________	____________________

II Fill in the blanks with the words or expressions given below. Change the form where necessary. Each word or expression can be used only once.

spearhead	initiate	holistic	optimize	with the exception of	be identified as
at all stages	on a voluntary basis	the upward trend	vie for	abide by	applicable

1. Barrette boots, which are made using silk satin, linen and leather in England over 1865—1875, ________________ attention with their combination of neutral shape and feminine decorative details, including a golden crossing band and bowknot on the toe cap.

2. We provide a space, equipment and training to our employees who participate ______________________.

3. The capital of Sichuan province, Chengdu, is helping to _________________ increased trade with countries involved in the Belt and Road Initiative through its advanced and integrated transport system.

4. China plans to partner with 76 WTO members to _________________ negotiations on

trade-related aspects of electronic commerce by building on existing WTO agreements, the Ministry of Commerce said.

5. We will strictly ________________ the laws on the protection of cultural relics and resolutely protect any relics excavated during construction.

6. The contract is also __________ to casual and seasonal workers with terms of less than a year.

7. More measures are expected to roll out to ___________ capital management for State-owned enterprises (SOEs) in China, as a part of the country's larger efforts to boost economic growth amid external challenges.

8. Island tourism holds substantial potential in China, but __________________ islands in Hainan—the present facilities need to be upgraded and the sophisticated planning necessary for the islands has yet to be developed.

9. English-language education should not be just teaching of the language, but a medium that can foster communication between cultures and promote ____________ development.

10. While perhaps religious in origin, Thanksgiving _____ now primarily _____________ a secular holiday.

11. Only extremely rigorous testing _____________ of development and production can guarantee impeccable quality.

12. The domestic tourism industry in China will continue to maintain an ________________ as the number of tourists increases and per-capita consumption capacity improves.

Ⅲ Complete the following statement by translating the Chinese in brackets into English.

1. Primarily made of proteins, silk is ______________________（在成分上与人体的皮肤接近）, making it extremely comfortable to wear.

2. Silk should be marketed ____（作为对环境友好的奢侈品）.

3. For instance, there is scope to expand ______________________________（丝绸织物用于室内装饰）.

4.（传统产业正在削减劳动密集型的丝绸生产）______________________________

____________________________, as urban industries lure farmers from a business in which incomes dropped radically in recent years.

5. At first restricted only to the emperor and his close relations, ____（丝绸象征着中国人对盛大典礼的喜爱）.

6. Widely used for clothing and decoration, ____________________________________（丝绸很快被投入商用）, becoming one of the principal elements of the Chinese economy.

7.（丝绸的价值以长度计算）____ and was equated with gold.

8. And to this day, 5,000 years after the first cocoon fell into the Empress' teacup, ____（中国仍然是首屈一指的养蚕大国和世界最大的丝绸生产国）.

Translation Skills

段落翻译（2）——段落的衔接与连贯

段落内的句子都是按照合理的、易于理解的逻辑关系连接成一个整体。这些关系既包含语法关系，又有语义和逻辑关系。以段落为单位进行汉译英，不仅要把各个句子译得正确和地道，还要考虑句与句之间的衔接，层次是否清晰，意思表达是否连贯，是否紧扣段落主题。请看下面各个译例：

（1）唐代文化是中国文化的一个高峰。尤其是古典诗歌在唐代发展到全胜时期。在唐代300余年的历史中，产生的流传于后世的诗歌就有48900多首。如此丰富的产品也使2300多位诗人在历史上留下了他们的名字。唐诗在创作方法上，现实主义与浪漫主义并举；在形式上有五言、七言绝句和律诗，还有优美整齐的近体诗。唐代最著名的诗人是李白和杜甫，他们都是具有世界声誉的诗人，后人将他们合称为“李杜”。

The Tang Dynasty witnessed a peak in Chinese culture. Especially for ancient poetry, <u>it</u> had <u>its</u> flowering in the Tang Dynasty. In more than 300 years of history in the Tang Dynasty, some 48,900 poems were handed down and remain widely known today. <u>So</u> many works also made more than 2,300 poets famous in history. <u>As far as</u> the writing technique is concerned, the Tang poetry combined realism and romanticism. In form, poetry of the Tang Dynasty contained four-line and eight-line verse with five or seven characters in each line. <u>Moreover</u>, “modern style” poetry, which is regular and polished, also appeared in the Tang Dynasty. The best-known poets of the Tang Dynasty were Li Bai and Du Fu, who are very prestigious in the whole world. Therefore, people of later generations have praised both of them as "Li Du" collaboratively.

该段译文中使用了大量的衔接手段，包括表示各种逻辑关系的衔接词如moreover，代词的使用如it，相同词汇的重复如poetry 和Tang Dynasty。

（2）近年来，在四川北部南坪县内，闪现出一颗五光十色的风光“宝石”，这就是人们赞不绝口的“神话世界”九寨沟。它镶嵌在松潘、南坪、平武三县接壤的群山之中，面积约6万公顷，距成都约400公里。九寨沟，由树正群海沟、则查洼沟、日则沟三条主沟组成，海拔平均在2500米左右。过去，沟中有9个藏族村寨，因此得名。

Jiuzhaigou(Nine-village Ravine), known as Fairyland, has shown its fascinating charm and beauty to millions of travelers in recent years. Located in Nanping county in the west of Sichuan Province, Jiuzhaigou is mounted in the hills along the borders of Songpan, Nanping and Pingwu counties, some 400 kilometers from Chengdu. It is made up of three ravines–Shuzhenquanhaigou, Zechawagou, and Rizegou, which combine to occupy about 60,000 hectares in area, and it is about 2,500 meters above the sea level. Its name "Nine-village Ravine" is supposed to be because there were nine Tibetan villages in the ravine in the past.

（大型画册《神话世界九寨沟》之序）

译文在信息的编排上做了灵活处理，第一句只保留了九寨沟如宝石般魅力无穷，吸引万千游客这一主要信息，因为这个就是整段主题所在，使得译文的主题句更加明确，其他如具体地理位置，则移到了下句，因为接下来各句分别从地理位置、组成部分、基本特征和得名由来四个方面来介绍九寨沟。因此原文中第二句中关于面积的信息被移到第三句中和海拔信息一起呈现。这样的处理使内容更为连贯，行文更为流畅。

Exercising Your Skills

Translate the following paragraphs into English and pay special attention to cohesion and coherence.

1. 复杂的汉字书写体系是中国古代文化遗留下来的瑰宝。而这一体系正面临着退化的命运。随着电脑和智能手机的迅速发展和普及，年轻人拿起笔却写不出字来的现象越来越常见。若不借助电子产品的帮助，很多人难以写出10万个日常用字。为此，中央电视台开播了一档《汉字听写大赛》节目，以引起人们对汉字的重视。

____

____

____

____

____

2. 西安是中国古代13个王朝的都城，毫无疑问，它是中国历史与文化的完美代表。西安居于“中国古都”之首，在中国历史上建都时间最长，影响力最大。它是丝绸之路（the Silk Road）的起点，是中华文明的发祥地。西安到处是令人惊叹的历史奇观，因此吸引着众多的国内外游客。那里有中国最古老、最壮观的博物馆和寺庙，其中最著名的是拥有2000年历史的兵马俑博物馆。

____

____

____

____

____

3. 中国历史早期，丝绸是被用来制作皇帝衣服的，但最终它被中国社会广泛采用。不管是做钓鱼线、造纸、制作乐器的弦，丝绸都很有价值。 在汉代，丝绸被作为皇室的礼品和贡品。它同黄金和钱一样，也成为一种普遍的交易物。中国农民用丝绸缴税。官员的薪水也以丝绸的形式支付。

____

____

____

____

____

Unit Project

In 2018, we commemorated the 40th anniversary of reform and opening-up of China. This is a time to celebrate. Over the past 40 years, China has grown into the world's second largest economy against the slowdown in world economic growth and the tensions affecting the world trading position in some sectors and regions; more than half of China's population has been lifted from poverty to a better life; and most importantly, China has not only witnessed fundamental changes at home, but also influenced global development, contributing Chinese wisdom to resolve global issues.

This is also an age of unprecedented scientific and technological progress. China continues to increase investment in R&D and in fields such as computer science, artificial intelligence as well as areas of bioscience. More Chinese companies such as Huawei and Alibaba are becoming leaders in their own fields. We can all see China's determination to become an "Innovation Nation."

Work in groups to complete the following tasks:

1. Work together to search the internet, library or other sources for achievements and progresses that have been made during the past 40 years in silk industry.

2. What roles have graduates of silk textiles, fashion, and other related majors played in the economic developments of our country; and what achievements do you expect to make in the future?

Unit 10 Keys

Glossary

Note:U=Unit, P=Passage

A

abound U1-P1
accentuate U4-P2
accessory U5-P1
acclaim U4-P2
accommodate U5-P2
accredit U5-P2
accumulate U10-P1
acquire U7-P2
adjoin U3-P1
aesthetics U7-P2
affluent U4-P2
afield U8-P2
alanine U2-P2
allegorical U1-P1
alliance U10-P2
allocation U8-P2
allure U6-P1
alternatively U5-P2
amity U8-P2
ammonium U8-P1
ancestor U4-P1
anecdotal U8-P1
antecedent U1-P2
antique U5-P1
antiquity U7-P2
apparel U2-P1
appealing U5-P1
appliance U6-P2
applicable U10-P2
appliqués U4-P2
apprentice U9-P1
apprenticeship U3-P1
arbitration U8-P2
arch U4-P1
archaeological U1-P1
archaeologist U1-P1
ardent U8-P1
arduous U3-P1
aristocratic U3-P1
aromatic U8-P1
artery U8-P1
artisan U1-P1
artisanal U4-P2
artistry U3-P2
assessment U6-P1
association U3-P1
assumption U6-P2
asymmetrical U4-P2
atelier U4-P2
attachment U10-P2
attire U6-P1
augur U6-P1
auspiciousness U3-P2
authentic U9-P2
authority U3-P1
avid U5-P1
azurite U5-P2

B

badge U4-P1
banner U1-P1

contemplate U9-P2
contemplation U7-P2
contributor U1-P2
contrive U7-P1
cooperation U6-P1
corridor U8-P2
costume U5-P1
counterpart U5-P2
courtesy U7-P2
couture U7-P2
couturier U4-P2
coveted U5-P2
craftsmanship U1-P1
crane U4-P1
crepe U3-P2
crepon U3-P2
crystallize U9-P2
cultivation U1-P2
cumulatively U8-P2
curation U7-P2
curve U6-P2
customization U7-P1
customize U7-P1

D

damask U1-P2
daze U1-P1
de facto U1-P1
debut U3-P1
decorate U6-P2
decorative U1-P1
dedicate U3-P1
deity U2-P1
delicate U5-P1
delineation U3-P1
demarcate U4-P1
denote U4-P1
density U7-P1
depict U5-P2
descendant U1-P1
detour U8-P2
devote U6-P1
diagonal U3-P2
dictate U1-P1
diffuse U5-P2
dignified U6-P2
dissemination U1-P1
dissolve U2-P2
distinguished U10-P1
diversified U6-P1
divine U2-P1
divinity U4-P1
documentary U7-P2
domestic U6-P1
don U4-P2
dowager U3-P1
downbeat U6-P1
downhill U6-P1
dramatic U6-P2
drawloom U1-P2
drenched U1-P1
duplicate U3-P1
durable U3-P2
dye U2-P2

E

echo U6-P1
ecumene U8-P2
edict U1-P2
edifice U6-P2
egret U5-P2
emanate U1-P1
embellishment U4-P2
emblematic U1-P1
embody U1-P1
embrace U7-P2
embroider U3-P1
embroidered U4-P1
embroidery U4-P1
emerald U4-P2
empirical U8-P1
emporium U6-P2

empress	U4-P1	filament	U2-P1
emulative	U7-P1	flagship	U6-P2
encompass	U8-P2	flair	U7-P1
enliven	U5-P2	floral	U4-P2
ensemble	U6-P2	floss	U9-P1
entail	U8-P2	flourish	U5-P2
entice	U8-P1	fluid	U2-P2
entrepreneur	U10-P1	folklore	U7-P2
entry	U6-P2	forum	U6-P1
envisage	U8-P2	four-shafted	U1-P2
envoy	U1-P1	fragmentation	U7-P2
epitome	U5-P1	freight	U7-P2
equate	U1-P1	fresco	U9-P2
escort	U5-P2	fret	U6-P1
essence	U9-P2	front	U6-P2
established	U7-P2	furnish	U6-P2
estate	U6-P2	fur-trimmed	U4-P2
eternity	U1-P1	fusion	U7-P2
evolve	U5-P2		
excavate	U8-P1	**G**	
excavation	U1-P1		
exceed	U10-P1	garment	U2-P1
exclusive	U3-P1	gauze	U3-P2
exclusively	U2-P1	gilt	U1-P1
execute	U6-P2	glamour	U6-P1
execution	U9-P1	gland	U2-P2
exquisite	U3-P1	glaring	U1-P1
extravagant	U9-P1	glistening	U1-P1
		glycine	U2-P2
F		gown	U3-P1
		grace	U6-P1
fabric	U1-P1	gratifying	U5-P1
façade	U6-P2	grind	U9-P2
facing	U1-P1	groove	U3-P1
facsimile	U9-P2	gross	U10-P1
fancy	U5-P1	gutta	U5-P2
feminine	U4-P2		
fertile	U2-P2	**H**	
fiddle	U6-P1		
fidelity	U4-P1	habotai	U3-P2
figurative	U4-P1	handful	U3-P1
figurine	U8-P1	harbour	U9-P2

hatch	U2-P1
haute	U4-P2
hegemony	U8-P2
heirloom	U9-P1
herald	U4-P1
heritage	U2-P1
heyday	U3-P2
high-end	U10-P1
holistic	U10-P2
hollow	U2-P2
horizontal	U3-P1
hue	U3-P1
humanity	U2-P1

I

illiterate	U9-P1
illuminate	U7-P2
imperial	U3-P1
inability	U6-P1
incubation	U2-P2
indicate	U7-P1
indulge	U5-P1
infrastructure	U8-P2
infused	U5-P1
inherit	U3-P1
initial	U1-P1
initiate	U10-P2
initiative	U7-P2
inlay	U5-P1
innovation	U6-P1
insignia	U4-P1
instar	U2-P2
instrument	U3-P1
intact	U2-P2
intangible	U2-P1
integrity	U5-P2
inter-operability	U10-P2
intersperse	U3-P1
intertwine	U8-P1
intimate	U3-P1
intricate	U3-P1
intrigue	U9-P2
involve	U5-P2
irrespective	U10-P2
Islam	U8-P1
Islamic	U8-P1
issue	U6-P1

K

Kesi	U1-P2

L

label	U6-P1
laborious	U2-P1
labour-intensive	U4-P2
landmark	U6-P2
lapis lazuli	U9-P2
larva	U2-P1
larval	U2-P2
lavish	U3-P1
leak	U1-P1
legacy	U3-P1
levitation	U8-P2
liken	U7-P2
limitation	U6-P1
lineage	U8-P2
literally	U9-P1
literary	U5-P1
longevity	U3-P1
loom	U1-P2
luminance	U1-P1
luxurious	U5-P2
luxury	U6-P1

M

made-to-order	U4-P2
magnificence	U3-P2
magnificent	U6-P1
magnolia	U7-P2
magpie	U4-P1

painstaking	U3-P1
panick	U1-P1
paperhanging	U3-P2
paradise	U6-P2
Parisian	U4-P2
pattern	U5-P1
peak	U6-P1
pen	U1-P1
pendant	U5-P1
perish	U8-P1
perpendicular	U1-P2
perpetuate	U7-P2
persecution	U1-P2
persevering	U9-P2
pharmaceutical	U10-P1
pheasant	U4-P1
pheromones	U2-P2
phoenix	U4-P1
pigment	U9-P2
pigment	U5-P2
pillar	U6-P1
poetic	U4-P1
polychrome	U3-P2
popularize	U10-P2
porcelain	U7-P2
porous	U1-P1
portal	U8-P1
portrait	U7-P1
pose	U6-P1
position	U6-P1
posterior	U2-P2
preaching	U5-P1
precede	U5-P2
precondition	U8-P2
predominantly	U8-P1
predominate	U3-P2
prestigious	U4-P2
prevail	U3-P1
prevalence	U7-P1
primitive	U2-P1
principal	U3-P1
procedure	U9-P1
process	U7-P1
procurement	U8-P2
profile	U5-P2
prompt	U8-P1
prop	U5-P1
property	U6-P2
prosper	U4-P2
prototype	U3-P1
punch	U1-P2
pupa	U2-P1

Q

quarter	U6-P1
quartz	U6-P2
quest	U1-P1
quilt	U2-P1

R

rayon	U2-P2
realignment	U8-P1
reappraisal	U7-P2
reckon	U6-P1
reclaim	U7-P2
reconciliation	U4-P1
reddish	U2-P2
reductive	U7-P2
reel	U1-P2
refine	U2-P1
refined	U9-P1
refreshingly	U8-P1
regime	U1-P2
reign	U1-P1
reignite	U5-P1
relapse	U8-P2
religiously	U7-P1
renaissance	U2-P2
rendering	U1-P1
replica	U3-P1
reposition	U7-P2
reputation	U6-P1

repute	U1-P1
resemble	U3-P1
reserve	U2-P1
reserved	U7-P1
resilience	U4-P1
resilient	U10-P1
respectively	U2-P2
resplendently	U1-P1
restoration	U6-P2
restore	U1-P2
retailer	U6-P2
retreat	U6-P2
revenue	U6-P1
revere	U2-P1
reverse	U9-P1
revival	U10-P1
revive	U10-P1
revocation	U1-P2
rheumatism	U9-P1
rib	U5-P1
ribbon	U2-P1
route	U1-P1
ruby	U4-P2
ruffle	U7-P2

S

saddle	U8-P1
salivary	U2-P2
Sanskrit	U8-P1
satin	U3-P2
savage	U7-P2
scheme	U6-P2
scripture	U1-P1
scroll	U5-P2
secluded	U10-P1
secrete	U2-P2
segment	U2-P2
self-interested	U1-P1
self-run	U6-P1
sericulture	U1-P1
serine	U2-P2
severity	U9-P1
sew	U3-P1
shimmer	U4-P2
shimmering	U1-P2
shortsighted	U1-P1
showcase	U9-P1
showroom	U6-P1
shrine	U8-P1
shuttle	U1-P2
silhouette	U4-P2
silicon	U7-P2
sink	U10-P1
skewer	U8-P1
slipcover	U6-P2
smuggle	U2-P2
solitary	U5-P1
soot	U5-P2
sophisticated	U9-P1
sought-after	U1-P1
souvenir	U6-P1
sovereignty	U8-P2
spearhead	U10-P2
spectacular	U4-P2
spin	U1-P2
spiracle	U2-P2
splendour	U9-P1
spool	U1-P2
spotlight	U4-P2
stagnate	U8-P2
stakeholder	U10-P2
staple	U6-P1
stark	U9-P2
statesperson	U7-P2
status	U1-P1
stave	U1-P2
steep	U4-P2
steeped	U7-P2
steer	U7-P1
strait	U8-P2
strand	U2-P1
strategic	U6-P1
strenuously	U7-P1

stretch	U5-P2
striking	U6-P2
string	U2-P2
stroke	U3-P1
stunning	U5-P1
subsidiary	U10-P1
substitute	U1-P2
surcoat	U4-P1
sustainability	U10-P2
sustainable	U10-P2
swamp	U9-P1
swelter	U9-P1
synonymous	U7-P2
synthesis	U2-P2
synthetic	U2-P2

T

tabby	U3-P2
tableful	U6-P2
tactful	U3-P1
tailor-made	U6-P2
tapestry	U1-P1
taut	U1-P1
telling	U1-P1
tender	U8-P2
testify	U1-P1
textile	U2-P1
texture	U2-P1
thriving	U10-P1
tranquillity	U7-P2
transcontinental	U1-P1
translucent	U10-P1
transmission	U8-P2
travelogue	U1-P1
traverse	U8-P1
treasured	U6-P1
tribute	U3-P2
triple	U10-P1
tycoon	U10-P1

U

ultimate	U3-P1
ultimately	U3-P2
undermine	U1-P1
undertake	U3-P2
undeterred	U1-P1
undo	U3-P1
undulate	U7-P2
unearth	U1-P1
uniform	U9-P1
unparalleled	U2-P1
unpick	U7-P2
unravel	U1-P2
unwind	U2-P1
upgrade	U6-P1
upmarket	U6-P1
upper-class	U6-P2
urbanization	U1-P2

V

vanish	U6-P2
velvet	U3-P2
vermilion	U5-P2
vibrant	U7-P1
vie	U10-P2
vigorous	U8-P1
viscose	U2-P2
voluntary	U10-P2
voluptuous	U10-P1

W

warp	U1-P2
warper	U1-P2
weave	U2-P1
weaving	U6-P1
weft	U3-P1
widening	U6-P2
witness	U6-P1
worldly	U3-P1

图书在版编目（CIP）数据

中国丝绸文化：英文 / 卢红君，陈茜主编. — 杭州 ：浙江大学出版社，2019.12

ISBN 978-7-308-19806-6

Ⅰ. ①中… Ⅱ. ①卢… ②陈… Ⅲ. ①丝绸—文化—中国—英文 Ⅳ. ①TS14-092

中国版本图书馆CIP数据核字（2019）第279443号

中国丝绸文化

卢红君 陈 茜 主编

责任编辑 李 晨
责任校对 郑成业
装帧设计 春天书装
出版发行 浙江大学出版社
（杭州市天目山路148号 邮政编码 310007）
（网址：http：//www.zjupress.com）
排 版 杭州林智广告有限公司
印 刷 广东虎彩云印刷有限公司绍兴分公司
开 本 787mm×1092mm 1/16
印 张 19
字 数 512千
版 印 次 2019年12月第1版 2019年12月第1次印刷
书 号 ISBN 978-7-308-19806-6
定 价 60.00元